ENVIRONMENTAL ECONOMICS FOR NON-ECONOMISTS

Techniques and Policies for Sustainable Development

2nd Edition

ENVIRONMENTAL ECONOMICS FOR NON-ECONOMISTS

Techniques and Policies for Sustainable Development

2nd Edition

John Asafu-Adjaye

The University of Queensland, Australia

 World Scientific

NEW JERSEY • LONDON • SINGAPORE • BEIJING • SHANGHAI • HONG KONG • TAIPEI • CHENNAI

Published by

World Scientific Publishing Co. Pte. Ltd.

5 Toh Tuck Link, Singapore 596224

USA office: 27 Warren Street, Suite 401-402, Hackensack, NJ 07601

UK office: 57 Shelton Street, Covent Garden, London WC2H 9HE

British Library Cataloguing-in-Publication Data
A catalogue record for this book is available from the British Library.

ENVIRONMENTAL ECONOMICS FOR NON-ECONOMISTS (2nd Edition)
Techniques and Policies for Sustainable Development

ISBN 981-256-123-4

Printed in Singapore by Mainland Press
This book is printed on acid-free paper.

CONTENTS

Preface

A second edition of this book has become necessary for two main reasons. Firstly, the book has established a broad readership base and has been well received, judging by the strong patronage over the last three years. I have received valuable feedback from various people which I have utilised in improving the presentation of the material. I would like to express my sincere gratitude to all those individuals who have provided reviews of the book, as well as the many readers who have made unsolicited comments. Secondly, a second edition offers an excellent opportunity to make a few editorial corrections, update the statistical information, and include additional material.

A number of changes have been made to enhance the usefulness of this book. Firstly, each chapter now contains a summary of the key terms and concepts used. This is meant to serve as a checklist for readers to ensure that they have obtained a firm grasp of the terminology and concepts. Secondly, solutions to all the numerical exercises have been provided at the end of the text. Thirdly, useful websites where the reader can access further information have been provided at the end of the book. This section would be particularly useful for readers engaged in research projects.

While the basic structure of the book has been retained, much of the material has been updated. The material in Chapter Two (Introduction to Ecological Economics) has essentially been retained. However, the discussion on Modelling Economy-Environment Interactions has been expanded to include the deficiencies of the input-output model, as well as information on the input-output econometric model. Given that one of the principal targets of this book is the non-economist who wants to know about environmental economics, Chapter Three (How Markets are Supposed to Work) retains its basic characteristic of trying to explain economic principles in simple terms. In the spirit of making the concepts comprehensible to the non-technician, the earlier material on indifference curves has been judged to be unnecessary and has therefore been removed. The discussion in Chapter

Four (Why Markets Fail) has been augmented with material on the topic of government (or policy) failure. To further elaborate on the solutions to the problem of market failure, an Appendix on the use of customary marine tenure systems in resource management has been added to this chapter.

In Chapter 10 (Economic Growth and the Environment), a section has been added on the relationship between economic growth and biodiversity loss. This issue has been virtually ignored in the current debate on the environmental effects of economic growth. Here, results from an empirical study are presented and discussed. The definitions of sustainable development (Chapter 11) have been expanded to include a wider range of perspectives. Finding appropriate indicators of welfare (or economic development) is essential in order to measure progress towards sustainable development. It is now widely accepted that the traditional measures (Gross Domestic Product and Gross National Product) are seriously flawed. In this regard, Chapter 11 contains a discussion on alternative measures of welfare that could be used to assess national progress. A new chapter (Chapter 12) has been added on Green Accounting and the Measurement of Genuine Saving. The discussion is complemented with case studies to illustrate the basic concepts.

This book is not only aimed at non-economists, but also economists who would like to learn about current environmental issues and possible solutions. As such, it tries to strike a balance between theory and applications. Although it is written from an economics perspective, the limits of analysing environmental problems using a purely economic analysis are acknowledged. For this reason, I have canvassed a multidisciplinary perspective, as evidenced by the inclusion of topics such as ecological economics, stakeholder analysis and multi-criteria analysis.

Once again, I would like to express my sincere thanks to those who have offered comments and suggestions for improving the book. I would also like to thank World Scientific Publishing Company, especially Joy Quek, for excellent work in all aspects of commissioning and providing general support for this and the previous edition.

John Asafu-Adjaye
July, 2004

1. Introduction

Within the past three decades, the world has witnessed a period of unprecedented economic growth. Total global output of goods and services increased from US$9.4 trillion to over US$25 trillion between 1960 and 1990 (UNDP, 1996). The benefits of this growth have not been evenly distributed. In 1993, the developed countries accounted for US$22.5 trillion of the total global gross domestic product (GDP) of US$27.7 trillion. Although some developing countries, especially those in South-East Asia, have shared in this growth spurt, others have missed out on the bonanza. Large parts of the developing world have been bypassed by the past three decades of economic growth. Since 1980, about 100 developing countries have experienced economic decline or stagnation; in 70 of these countries, average incomes in the late 1990s were below the 1980s (UNDP, 1997).

The impressive performance of the world economy has come about mainly as a result of globalisation. 'Globalisation' is a term that was coined in the 1990s to refer to the integration of the global economy brought about by the rapid developments in information technology and the reduction of international trade barriers. Globalisation has created a near 'borderless' world and has facilitated free trade and flows of private capital between countries. Global trade increased from US$4,345 billion to US$6,255 billion between 1990 and 1995. Transfers of net private capital into low-income and middle-income countries amounted to US$180 billion in 1995, compared to official development assistance of US$64 billion (World Bank, 1999).

The growth of the global economy has brought with it several benefits such as improvement in health and living conditions in many developing countries. For example, in many developing countries, infant mortality rates have declined, life expectancy has increased and illiteracy rates have declined over the past three decades. However, disparities in poverty and income distribution persist between regions and within countries. Absolute poverty in parts of Africa, Latin America and the Caribbean has increased, and the gap between the developed and developing countries has widened.

Economic growth is required to meet the needs of a growing population. However, rapid growth has serious implications for our physical environment. Expansion of agricultural land is essential to produce more food.[1] Activities such as land clearing and the use of pesticides have potential adverse environmental impacts. Industrial production is required to house, clothe and feed the population. However, some industrial processes result in the production of air and water pollution, as well as the generation of toxic waste products.

Energy is a vital input to transportation, industrial production and agricultural production. It also provides other important domestic services such as heating, cooling and lighting. At the present time, the developed countries account for about 70 percent of carbon dioxide (CO_2) emissions even though they account for less than 20 percent of global population (UN, 1997). Energy demand is projected to increase rapidly in the developing countries. It is estimated that developing country share of world energy demand will increase by almost 40 percent by 2010. This demand on world energy resources will come about as a result of rapid economic expansion, especially in the South-East Asian region (IEA, 1996).

The unrestrained use of fossil fuels poses a serious threat to the environment. There is the potential to increase greenhouse gas emissions and global warming. Although the precise impacts of climatic change are not quite clear, some possible outcomes have been identified. If current trends of energy use continue, the average global temperature is expected to increase by 1.0°C to 3.5°C over the next century (WHO, 1996). There will be rise in the sea level of about 30cm; there will be accummulation of ice and snow in polar ice caps; and there will be severe storms, drought and flooding due to the climatic changes. There could also be an increase in insect-borne diseases such as malaria, and some animal and plant species could become extinct.

1.1   The Role of Environmental Economics

This book is concerned with the application of economics to the solution of global and local environmental problems. 'Economics' may be defined as a study of choice given limited financial and natural resources. In a world of unlimited resources, the choice an individual or society makes has no implications whatsoever. However, in view of the finiteness of resources,

---

[1] Of course, with improved technology more food could be produced without necessarily expanding agricultural land. However, the fact of the matter is that most developing countries do not utilise high technology in agriculture.

every possible choice has an associated cost. We refer to this in economics as **opportunity cost**. This term is defined later. Consider a situation in which the government wants to construct, say, an airport on pristine land. In this case, given limited funds and natural resources, a decision to go ahead with the project precludes the use of the land for other purposes.

As will be explained in the next chapter, the economy is a complex system and when modelling such a system simplifying assumptions need to be made. Due to the wide range of issues relating to economic systems, various specializations have arisen within the economics profession. **Microeconomics** is the study of the economy at the individual or firm level, whereas **macroeconomics** is the study of the economy at the aggregate level. Examples of the former include the behaviour of economic agents (consumers and producers) and effects on demand conditions and prices, whereas examples of the latter include issues such as changes in employment (or unemployment), inflation, savings, investment, and so on. The subdiscipline of **econometrics** uses economic concepts and statistical methods to carry out quantitative analyses of economic issues.

Traditionally, the study of **natural resource economics** was concerned with the application of economic theory and quantitative methods to determine the optimum allocation and distribution of natural resources. However, with the rise of environmental concerns in the 1960s, **environmental economics** has evolved as a subdiciptine of economics which not only includes aspects of natural resource economics (e.g., allocation and distribution of resources) but also broader issues such as the interactions between the economy and the environment. Environmental economics also deals with institutional and ethical issues associated with the conservation and use of natural resources. Tisdell (1993:3) defines environmental economics as the 'study of the impact of economic activity on the environment as well as the influence of the environment on economic activity and human welfare'. This broad definition includes man-made environments such as built (urban) environments, historical and cultural environments. Within the last decade or so, **ecological economics** has emerged as a new subdiscipline of environmental economics. Ecological economics emphasises the constraints that the natural ecosystem places on the economic system. The subject of ecological economics is discussed at length in Chapter 2.

1.2   Defining the Natural Environment

In view of the fact that the environment is a major focus of this book, it would be useful to define the natural environment in order to set the discussion in an appropriate context. Broadly speaking, the natural environment comprises two types of resources: renewable resources and non-renewable resources.

As the name suggests, renewable natural resources are biological resources that have a capacity for regeneration. Examples are forests, animals and micro-organisms. In theory, renewable resources have the capacity to provide infinite services. However, we demonstrate in the next chapter that there are some ecological constraints to this possibility.

Non-renewable resources are finite in terms of supply. There are three major types of non-renewable resources: exhaustible resources, recyclable resources and non-renewable resources with renewable service flows. Examples of exhaustible resources include coal, crude oil and bauxite. Examples of recyclable resources include most metals such as tin, copper, aluminium and gold. Examples of non-renewable resources with renewable service flows include land, seas and rivers.

It is important to emphasise that the above classification is not static. A renewable resource can become non-renewable if poorly managed. For example, indiscriminate fishing could reduce the population to a level where the species cannot reproduce. A piece of land is a finite non-renewable resource that could be used to provide a renewable service such as cultivation of crops.

The approach adopted in this book is to consider resources not in isolation but as a system—the ecosystem. In this regard, interactions within the system are important. For example, although a stand of forest timber would be valued for its timber in the traditional economic approach, the approach taken here would be to also consider the contribution of the biological functions of the forest cover, the wildlife, the biodiversity functions, and so on.

1.3   Overview of this Book

The material is presented in three parts. Part I presents conceptual issues that are required to analyse how the economic activities impact on the environment. Part I leads off, in Chapter 2, with an introduction to the relatively new subdiscipline of ecological economics. The adjective 'new' is used here because even though some of the ideas have been around since the the 18th Century, they are only now being applied in the area of environmental economics. Many people would agree that the concept of free markets does not work well for environmental resources. In Chapter 3, we demonstrate how markets are supposed to work under traditional (neoclassical) economic theory. We then go on to explain, in Chapter 4, why markets fail to work the way they should in the case of environmental resources.

Part II of the book presents various tools for environmental policy analysis. It begins with techniques for valuing environmental damage and benefits. In recent years, as environmental issues have grown in importance, governments have been forced to legislate laws protecting the environment. Measures of environmental damage are now sought to assess penalties and to determine compensation levels in litigation cases. In many countries, the law requires project developers to conduct an environmental impact assessment and part of this process requires an estimate of the amount and value of any potential damage. Chapter 5 discusses recently introduced techniques that could be used to estimate the value of environmental damage and benefits. A consistent framework is required in development planning and policy analysis. Chapter 6 introduces the methodology of cost-benefit analysis (CBA), with particular emphasis on public projects. Cost-benefit analysis is not always adequate, or even appropriate, in certain situations. Therefore, Chapter 7 introduces additional methods—cost-effectiveness analysis (CEA), impact analysis (IA) and stakeholder analysis (SA)—that could be used to complement a CBA. Techniques such as CBA and CEA are designed for decisions with single objectives. To complement these approaches, Chapter 8 introduces Multi-Criteria Analysis (MCA) which is applicable to decisions involving multiple objectives that may be conflicting or competing.

All forms of environmental degradation, whether local or regional, have global implications in the long term. Part III of the book examines global environmental issues. Chapter 9 discusses the effects of population growth and resource use on the environment and the policy implications. Chapter 10 reviews the debate on the relationship between economic growth and the environment. This chapter also includes the effects of economic growth on biodiversity. Chapter 11 adresses the issue of sustainable development.

'Sustainable development' is, perhaps, the most widely used term in both governmental and non-governmental organisations. However, it could be among the least understood terms in use today. In this chapter, definitions from various perspectives are presented. Practical issues such as measurement of sustainable development and implementation constraints are discussed. Chapter 12 delves deeper into practical tools that could be used to assess progress towards sustainable development. Finally, Chapter 13 concludes with an assessment of current global environmental trends and their policy implications.

References

International Energy Agency, IEA, (1996). *World Energy Outlook 1996.* Organisation for Economic Co-Operation and Development, Paris.

Tisdell, C. (1993). *Environmental Economics: Policies for Environmental Management and Sustainable Development.* Edward Elgar, Aldershot, UK.

United Nations, UN (1997). *Critical Trends: Global Change and Sustainable Development.* United Nations, New York.

United Nations Development Programme, UNDP (1997). *Human Development Report.* Oxford University Press, New York and Oxford.

United Nations Development Programme, UNDP (1996). *Human Development Report.* Oxford University Press, New York and Oxford.

World Bank (1999) *World Development Indicators 1999CD-ROM.* World Bank, Washington, D.C.

World Health Organization, WHO, (1996). *Climate Change and Human Health.* World Health Organization, Geneva.

Part I. Introduction to Environmental Economics: Theoretical Foundations

2. Incorporating the Environment into the Economic System: Introduction to Ecological Economics

Objectives

After studying this chapter you should be in a position to

- explain the basic functioning of the traditional economic system and its limitations as far as the environment is concerned;

- explain the relationships between the economic and environmental systems;

- explain the laws of thermodynamics and their implications for the economy-environment system; and

- describe various approaches to incorporating environmental concerns into economic models and their limitations

2.1 Introduction

As concern for the environment has increased in the past few decades, so has the need for sustainable development or 'sustainability' policies. In general, many decisions relating to development policy have been determined on the basis of economics. However, we argue later in this chapter that traditional economic models have tended to ignore the role of the environment. In order to make effective plans for sustainable development, there is the need to consider the interactions of the environmental and economic systems. In this regard, the purpose of this chapter is to demonstrate the limitations of the traditional economic model and to introduce the reader to the relatively new discipline of ecological economics which attempts to link the environment to the economy.

The chapter is divided into six sections. The next section presents the traditional economic system and discusses its main limitations. Section 2 defines the meaning of 'ecological economics'. Section 3 describes the economy-environment system, while Section 4 introduces the laws of thermodynamics. Section 5 considers approaches to modelling environment-economy interactions, while the final section contains the summary.

2.2   What is Ecological Economics?

Ecological or 'green' economics means different things to different people. It can be defined from various perspectives such as biology, chemistry, physics, engineering, mathematics, sociology, politics and economics. Before we say what ecological economics is, it may be helpful to first say what it is not. Ecological economics is not synonymous with environmental economics or natural resources economics, although both ecological economics and natural resource economics are subsets of environmental economics. 'Ecology' can be defined as the science of the self-organisation of nature (Faber *et al.,* 1996). 'Nature' or the environment is a broad concept that encompasses the universe—plants, animals, ecosystems, materials and humans. Webster's New World Dictionary defines 'ecology' as 'the branch of biology that deals with the relations between living organisms and their environment'.[2] According to Costanza *et al.* (1991), ecological economics sees the human economy as part of a larger whole. Its domain is the entire web of interactions between economic and ecological sectors.

Ecological economics is therefore concerned with how environmental (or ecological) and economic systems interact. On the other hand, natural resource economics is mostly concerned with the best way of exploiting renewable and non-renewable resources. The main difference between ecological economics and natural resource economics is that, in addition to looking at the exploitation of resources, the former also considers social and ethical issues, as well as placing emphasis on ecological processes.

---

[2] The Greek word 'oikos' is the common root for the 'eco' in both economics and ecology. Oikos means 'household' and so it could be argued that ecology is the study of nature's housekeeping, while economics is the study of human housekeeping.

To give an example of how ecological economics and natural resource economics differ, consider a natural resource such as a tropical rain forest. Typically, natural resource economics will look at the optimum rate of harvest to achieve objectives such as maximum sustained yield, given parameters such as stumpage prices and interest rates.[3] On the other hand, ecological economics will consider the issue of how exploitation affects the rights of future generations, as well as the rights of other forms of life in the ecosystem.

Ecological economics also differs from traditional or neoclassical economics in several ways. For example, neoclassical economics is based on the assumption of 'rational' economic behaviour based on utility maximisation or profit maximisation.[4] Although neoclassical economists view environmental problems as an externality problem, the 'purists' amongst them would recommend limited government intervention. For example, they would advocate that all the government needs to do is to, say, tax the negative externality and allow market forces to deal with the problem of allocation. However, ecological economists would say that neoclassical economics provides only a partial view of a complex problem and as such ecological factors should be included in the analysis of such problems.

A key concept in ecological economics is the concept of **evolution**. This concept can also be interpreted from different perspectives. Thus, for example, one can speak of geological, biological, social, political and economic evolution. Faber *et al.* (1996) define evolution as the process of the changing of something over time. In the biological sense, evolution of a species can be described as change in the gene pool that is common to a group of organisms belonging to the same species.[5] In general, evolution or evolutionary processes enable a species to adapt to its environment through the selective replacement of weaker individuals in the population. This process of replacement of individuals (also referred to as **natural selection**) ensures that the fitter individuals in the population make a genetic contribution to future generations. A number of prominent economists have attempted to extend the concept of evolution in biology to the economic

---

[3] Maximum sustained yield is the largest possible average yield of wood sustainable over an indefinite period. Stumpage price is the sale price of logs.

[4] The basic foundations of neoclassical economic theory are discussed in some detail in Chapter 3.

[5] Mayr defines a 'species' as 'a group of actually or potentially interbreeding populations that are reproductively isolated from other such groups' (Mayr, 1942:120).

system. For example, Norgaad (1984) has used the concept to explain environment-economy interactions as processes governed by feedback and learning.[6]

The idea of a relationship between the economy and the environment is not new. In his seminal book, *Principles of Economics*, the economist Alfred Marshall drew parallels between economics and biology (Marshall, 1930). However, Marshall's views on economics and biology were never taken seriously by economists until Boulding (1966) resurrected the issue with his concept of the **'Spaceship Earth'**. He used the 'Spaceship Earth' to make the point that human beings live in a closed system, the earth, and are dependent on it for sustenance. The earth does not receive anything from the outside except the sun's energy. Other forms of energy must be produced from the resources available to it and the same system must also absorb the waste products generated by consumption and production activities.

Ecological-economic models began to emerge about three decades ago in response to the limitations of traditional economics in tackling environmental problems. The first people to present a systematic framework for integrating economic and ecological systems were Ayres and Kneese, with their concept of **'materials balance'** (Ayres and Kneese, 1969; Ayres *et al.* 1970). The basic foundations of the materials balance model is the Conservation of Mass Principle, borrowed from the First Law of Thermodynamics. This approach is discussed in a little detail later. Hannon (1986, 1991) attempted to link ecological theory to economic behaviour and the price system, using input-output analysis. The input-output framework was extended to include the emission of waste residuals (e.g., James 1993). Crocker and Tschirhart (1992) and Crocker (1995) attempted to include ecological functions such as the food chain into a general equilibrium model. In recent years, neoclassical growth theory has also been extended to incorporate environmental issues such as sustainability[7] (Toman *et al.* 1994) and global warming (Nordhaus, 1990, 1993). Attempts have also been made to account for the environmental impacts of international trade (Barbier and Rauscher, 1994). Finally, some researchers have attempted to incorporate environmental concerns using systems analysis and systems dynamics (e.g., see Bergman, 1991; van den Berg 1993; van den Berg and Nijkamp, 1994). Some of these models are discussed later in this chapter.

---

[6] See, other contributions by Boulding (1981), Nelson and Winter (1982) and Clark and Juma (1987).

[7] The issue of 'sustainability' is discussed in Chapter 11.

A major obstacle impeding progress in attempts to incorporate ecological functions into economic models is how to value the goods and services provided by the ecosystem in monetary terms. This problem arises because the common denominator in economic models is price and most often ecosystem 'goods' and services are not traded in markets and therefore have no prices. Recently, methodological advances have been made that allow such goods to be valued. These methods are discussed in some detail in Chapter 5 of this book.

To summarise the discussion so far, it must be emphasised that ecological economics is a sub-discipline of environmental economics. The main thrust of ecological economics is how the ecosystem interacts with the economic system. Ecological economics attempts to model these interrelationships in order to draw conclusions for policy making. Ecological economics cuts across a wider domain than neoclassical economics, embracing such diverse fields as the natural sciences, philosophy, political science and sociology. In the following sections, we introduce simplified representations of an economic system, an ecosystem and an economy-environment system.

2.3   Economy-Environment Systems

A 'system' comprises a collection of objects or entities that are bounded in terms of space and time. The entities interact with each other through various 'processes'. There are three types of systems: **isolated**, **closed** and **open** systems. Given that we will discuss thermodynamics later, it is useful to define these types of systems in terms of their thermodynamic properties.

- In an isolated system neither energy nor matter is exchanged with the surrounding environment;
- In a closed system energy is exchanged with the surrounding environment but not matter, and
- In an open system both energy and matter are exchanged with the surrounding environment.

In the following sections, we consider a traditional economic system, an ecosystem and an economic-environment system.

2.3.1   A Traditional Economic System

A traditional economic system comprises producers, consumers and markets (Figure 2.1). Firms produce goods and services, using inputs of capital and labour supplied by consumers. These goods and services are offered for sale in the market and are purchased by consumers. Consumers who supply labour and capital to firms are rewarded with wages, profits, interest or rent. This simple model can be extended by including a government sector that controls the market by setting rules and regulations. The economic model presented below is a closed system in the sense that its boundaries are limited to consumption, production and exchange between economic agents. It completely ignores the flow of materials and energy that cross the boundaries of the system. Activities or resources that are unpriced are not considered to be of value in the economic system. For example, the harvesting and sale of plants for food will be considered as a valuable activity in the economic system. However, the production of complex organic molecules in plant material will be ignored. The main reason for this anomaly is that markets do not exist for the complex organic molecules used in the production of plant material.

2.3.2   An Ecosystem

An example of an open system is the ecosystem. The 'ecosystem' can be defined as the environment in which organisms (including humans) live. The environment, in this case, includes both the physical (abiotic) and the biological (biotic) conditions in which the organism lives. Figure 2.2 depicts a simplified representation of an ecosystem. The main characteristics are the flow of low-entropy energy from the sun into the ecosytem. The organisms in the ecosystem capture and transform this energy, combining it with other raw materials such as water and CO_2 to provide for the growth, maintenance and reproduction of the species. The conversion of low-entropy energy into other forms of energy (e.g., heat) is not 100 percent efficient and therefore there is release of high-entropy energy or waste back into the ecosystem.

An important feature of the ecosystem not shown in Figure 2.2 is 'feedback'. Feedback processes are means by which the various components of the ecosystem interact and achieve a state of equilibrium. There are two types of feedback processes: positive feedback and negative feedback.

In a positive feedback the eventual response of the species is in the same direction as the initial stimulus, whereas in negative feedback the response is in an opposite direction. To give an example of negative feedback, consider a predator-prey relationship in an ecosystem.

Figure 2.1 A traditional economic system

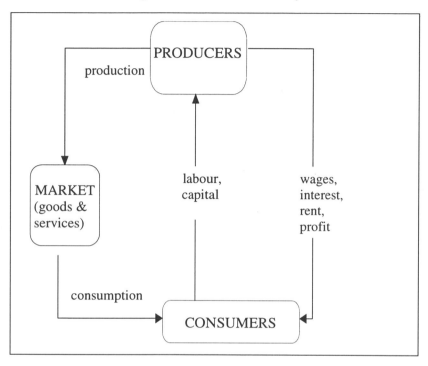

An increase in the density of the prey species initially stimulates an increase in the density of the predator. However, over time, the increase in the density of the predator will cause the density of the prey to decline. Thus, in this case, negative feedback causes equilibrium to be achieved in the ecosystem.

Ecosystems may be broadly classified into two main components: **autotrophs** and **heterotrophs**. Autotrophs are green plants that use the sun's energy to build complex organic molecules from simple inorganic molecules. On the other hand, heterotrophs are higher order organisms that feed on autotrophs in order to obtain energy. A good example of heterotrophs is the human species which consumes agricultural products in order to gain the necessary energy reserves to provide labour inputs for production activities within the economic system. Once again, it must be pointed out that the conversion of food into other forms of energy is not 100 percent efficient. As such, as depicted in Figure 2.2, high-entropy energy or waste is released into the ecosystem.

Figure 2.2 A schematic representation of an ecosystem

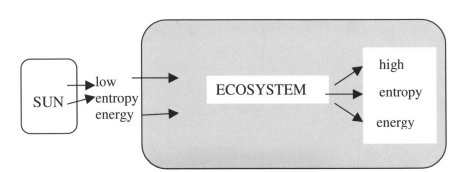

To conclude this brief discussion on the ecosystem, a couple of points about the ecosystem are worth noting. First, the 'productivity' of the ecosystem is dependent on how efficient the species is in capturing and transforming energy and other raw materials for maintenance, growth and reproduction. Second, 'equilibrium' in the ecosystem is not static. As a result of feedback, ecosystems move their equilibrium position over time and changes occur in the composition and abundance of the species. These changes form part of the evolutionary processes (alluded to earlier) that continually occur in ecosystems.

2.3.3   An Economy-Environment System

An alternative portrayal of the economic system is an open system where there is interaction with a distinct environmental system. Both the economy and the environment are open sub-systems of a larger system, the universe.[8] Such a system is depicted in Figure 2.3.

---

[8] Note that this broad definition that includes the universe is, in effect, a closed system.

Figure 2.3 An economy-environment system

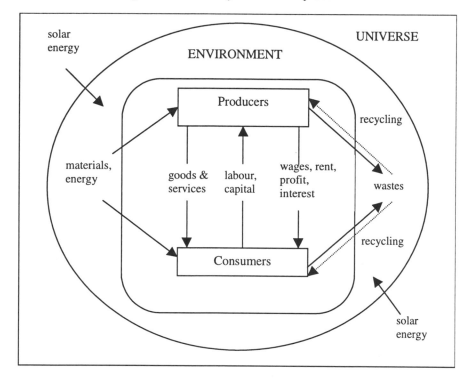

Firms produce goods and services using raw materials such as minerals, agricultural products, timber, fuels, water and oxygen that are extracted from the environment. These goods are sold in the market as either consumer goods or as intermediate goods for the production process.

Nearly all the material inputs to the production and consumption processes are returned to the environment as waste. The waste products are mainly in the form of gases (e.g., carbon monoxide, carbon dioxide, nitrogen dioxide, sulphur dioxide), dry solids (e.g., rubbish and scrap), or wet solids (e.g., wastewater). Both solid and liquid waste products from the household and production sectors may go through a further processing stage before being returned to the environment as waste. However, as we shall see later, processing only changes the form and ultimate destination of the residual flows. Consequently, the total amount of materials returned to the environment will remain unchanged. This approach to viewing economy-environment interactions is based on the principle of 'materials balance' that we will discuss later under the laws of thermodynamics.

The environment in the above system can be seen as playing four important roles:

- As a provider of energy and raw materials inputs to producers and consumers;
- As a receptacle for the waste products of producers and consumers;
- As a provider of amenities to consumers (e.g., recreation); and
- As a provider of basic life-support functions for humans.

Two important points are worth noting about the economy-environment system shown above. First, there is a strong interrelationship between the three types of support provided by the environment. For example, there is a limit to the extent of the environment's capacity to assimilate waste. Pollution and environmental degradation begin to occur when this assimilative capacity is exceeded. Furthermore, once this limit is exceeded the ability of the environment to provide other services (e.g., provide inputs) is compromised. Second, we need to view the natural environment not only as a resource but also as an asset similar to traditional assets such as land, labour and capital. The value of this resource must therefore be integrated into the economic system. In traditional accounting practice, the depreciation of capital assets is taken into account when assessing financial performance. The same consideration must be given to environmental assets when analysing economic performance. That is, we need to ensure the maintenance of the quality of the natural environment in the same way that we would maintain fixed assets such as plant and equipment.

2.4   Thermodynamics and the Environment

Thermodynamics can be described as the science of energy. Energy can be defined as the potential to do work or supply heat. Work is involved whenever matter is changed in terms of physical or chemical structure, or in terms of location. The two major laws of physics are the First and Second Laws of Thermodynamics. In this section, we briefly outline these laws and consider their implications for the economy-environment system. The First

Law is also known as the **Law of Conservation of Mass and Energy**, whereas the Second Law is often referred to as the **Entropy Law**. In this section, we provide a brief sketch of the historical origin of these laws.

2.4.1 The First Law of Thermodynamics

The science of thermodynamics originated in the engineering discipline as a result of efforts to understand the functioning of heat engines during the 18th Century. The French engineer Sadi Carnot was the first person to analyse how heat could be transformed into mechanical work using a heat engine. He made comparisons between a heat engine and a water wheel at a mill: as the water produces work by flowing from a high to a low elevation, so does heat produce work by 'flowing' from a high to a low temperature within a heat engine. Although Carnot himself did not use the word 'entropy', the Entropy Law, which is defined below, is believed to have originated from his observation that the potential amount of work that heat can produce depends only on the difference in the temperatures of two entities between which heat is exchanged.

Carnot's observation about the equivalence between work and heat as different forms of energy was confirmed several years later by both theoretical and experimental physicists. This led to the establishment of the principle of conservation of mass and energy or **'The First Law of Thermodynamics'**. According to this law, energy cannot be created or destroyed, although it can be transformed into different forms such as heat, chemical energy, electrical energy, kinetic energy, and so on. The First Law is also referred to as the law of conservation of mass and energy because it implies that, although the form of energy may change, the total amount of energy in the system remains constant. The First Law applies to a closed system in which only energy crosses the boundaries (e.g., see Figure 2.2).

To give an example of the First Law, consider the burning of firewood to provide heat in an insulated room. After all the wood has been burnt, the chemical energy in the wood would have been transformed into a higher room temperature. Because the air temperature would have increased by the same amount as the decline in the energy content of the firewood, the total energy in the room would be constant. The main implication of the First Law is that raw materials cannot be consumed or used up after they have been extracted from the environment.

2.4.2   The Second Law of Thermodynamics

Although it was Carnot who introduced the notion of 'entropy', Rudolph Clausius is credited with formalising the concept in 1854. Before stating the Second Law, it is useful to make a distinction between work and heat. Work, including all other forms of energy except heat, can be converted into heat completely. However, the converse (i.e., conversion of heat into work) cannot occur with 100% efficiency. The reason is that the process of converting heat into work entails loss of heat. This observation forms the basis of the **Second Law of Thermodynamics**, or the **Entropy Law** which states that in an isolated and closed system entropy always increases or, in reversible processes, remains constant. This implies that heat will always flow from a hotter to a colder body, and that heat cannot be transformed into work with 100% efficiency. Another way of describing this phenomenon is to say that an isolated system reaches a state of internal equilibrium when the entropy reaches a maximum level. Clausius used this property of systems to explain why different types of gradients such as temperature, pressure, and density tend to level off or disappear over time.

Entropy can be defined as the amount of energy available for work. It can be used as a measure of the quality of heat in the sense that low-entropy raw material is 'more useful', but high-entropy waste material is 'less useful'. Ayres (1998) has argued that this definition of entropy could be misleading and that a more useful term is **'exergy'**. Exergy can be defined as the 'potential work that can be extracted from a system by reversible processes as the system equilibrates with its surroundings' (Ayres, 1998:192). Exergy can therefore be described as the 'more useful' part of energy. Thus, for example, work has 100% exergy whereas heat has much less exergy. There are four types of exergy:

- kinetic exergy, which is associated with relative motion,
- potential field exergy, which is associated with gravity or electromagnetic field gradients,
- physical exergy, which is associated with pressure and temperature gradients, and
- chemical exergy, which is associated with chemical gradients.

The laws of thermodynamics can be re-cast in terms of exergy. Thus, regarding the First Law, we can say that in an isolated system energy consists of exergy and unavailable energy and the sum of the two remains constant within the system. In terms of the Second Law, we can say that the exergy of an isolated system decreases over time.

2.4.3 Interpretations of the Second Law of Thermodynamics

The concept of entropy has led to the development of other concepts about nature and natural systems. For instance, the property that entropy always increases in an isolated system led Sir Arthur Eddington, an astronomer and scientist, to introduce the notion of 'time' in describing a system. He referred to it as **'The First Arrow of Time'** (Layzer, 1976). According to Eddington, the direction of time is the direction in which entropy increases. 'The First Arrow of Time' concept implies that time is irreversible. This is in contradiction to Sir Isaac Newton's Laws of Motion which imply that time is reversible. Newton's laws are assumed to hold even when the direction of motion of a body (or bodies) is reversed. That is, in the absence of energy losses, the motion of bodies can be described as symmetrical in the sense that a forward motion is equivalent to a backward motion. An implication of time reversibility is that the future is merely a continuation of the past, and that change or evolution does not occur. This view is embodied in the concept of 'Laplace's Demon' which states that, given the present positions and velocities of all particles in the universe, one can infer the past and predict the future (Prigogine and Stengers, 1984). In general, neoclassical economic models appear to have adopted the Newtonian view of time.

Another outcome of the Entropy Law is the concept of 'order' in systems. For example, as already indicated above, heat will flow from a hot to a cold body. However, the reverse can never occur. This implies that a process such as temperature gradient equalisation is irreversible in time. Many scientists interpret this property to mean that a system has the tendency to increase in 'disorder' due to increase in entropy. Clausius argued that because heat flows from hot to cold bodies, differences in temperature and concentration of matter in all bodies in the universe (i.e., planets, stars and galaxies) will eventually level out, resulting in the maximisation of entropy in the universe. He referred to this as the **'heat death'** of the universe.

The accuracy of these predictions remains to be tested. Strictly speaking, our planet is not a closed system because we receive energy from the sun. However, solar energy can be considered as a major constraint to the flow of sustainable energy. Therefore, in the long run, economic growth will be limited by solar energy and our ability to convert it to work.

A third interpretation of the Second Law of Thermodynamics has been in terms of its relationship to biological systems. Simple life forms such as algae are hypothesised to have developed out of basic molecular structures, from which even more complex structures evolved. This phenomenon in which a system tends to develop a more complex organisational structure has been referred to as **'The Second Arrow of Time'**. The Second Arrow of Time also implies that time is irreversible because there is a distinction between the past and the future. During the 19th Century, it was felt that the Second Arrow of Time concept contradicted the Second Law because self-organisation implies an increase in 'order' (i.e., decrease in entropy) rather than an increase in 'disorder' (i.e., increase in entropy). However, in the early 1940s, Erwin Schrödinger, a physicist, explained that living organisms operated within open systems in which there is exchange of matter and energy with their immediate environments. He went on to argue that in open systems that are far from thermodynamic equilibrium, entropy could decrease through the importation of low entropy from the surrounding environment and export of high entropy (Schrödinger, 1944).

The concept of the second arrow of time has been used to describe how economic systems evolve over time by means of capital goods, institutional structures and technical progress. This particular aspect of self-organisation in economic systems has led to the development of a new discipline in economics referred to as **evolutionary economics** (Dosi and Nelson, 1994). Brooks and Wiley (1988) have undertaken theoretical work to explain evolutionary processes using models of thermodynamics.

2.4.4 Implications of the Laws of Thermodynamics for the Economy-Environment System

The two laws of thermodynamics have, on the basis of both theoretical and empirical evidence, been proven to hold consistently. Economic activities such as the production and consumption of goods and services require energy as a major input. At issue is whether the economic system is also subject to the laws of thermodynamics. If the economic system is considered as a sub-system of a larger system, the universe, which itself is a closed system (see Figure 2.3), then it can be argued that the economic system must be subject to the laws of thermodynamics. We consider below, the implications of the laws of thermodynamics.

Implications of the First Law

The First Law of Thermodynamics has two important implications for the economy-environment system. These implications can be considered under two headings: conservation of energy and conservation of mass (the mass balance principle). Under conservation of energy, the law implies that energy inputs must equal energy outputs for any transformation process because energy cannot be consumed or used up. For mass conservation, the law implies that mass inputs must equal mass outputs for every process. This implies that any raw material inputs used in the production and consumption process must eventually be returned to the environment as high-entropy waste products or pollutants. Recycling can help to reduce the amount of waste, to some extent. However, as indicated above, recycling cannot be 100% effective and therefore cannot fully convert the unavailable energy to work.

In the next chapter, we will discuss how the market system works in neoclassical economics and how efficient allocation of goods and services is achieved through the price system. This is followed, in Chapter 4, by a discussion of 'externalities' or 'market failure'. Externalities occur because the economic system fails to recognise the value of goods that are not sold in the market place (i.e., non-market goods). Most environmental goods fall under this category of goods. The outcome of externalities or market failure is inefficient allocation of resources. Thus, for example, excess pollution is produced because the cost of pollution is not included in the production (or consumption) decisions of economic agents. Neoclassical economics prescribes policies such as taxes to 'internalise' externalities in an attempt to create incentives for reducing externalities in the economic system. This has been referred to as *a posteriori* correction for externalities (Ruth, 1993). It has been suggested that in an economy-environment model that takes account of the materials balance principle, the notion of 'externalities' does not exist. This is because all the factors in both the economic and environmental sub-systems (e.g., material and energy flows) are accounted for.

Implications of the Second Law

The Second Law implies that economic processes (i.e., production and consumption) are time irreversible in the sense that waste material cannot be fully converted to useful energy. This has led some economists such as Daly (1991) to propose a transition to a 'steady state' economy. A steady state economy is one in which there are constant stocks of people and physical wealth that are kept at a desired level. Maintenance of a steady state implies that:

- use of (conditionally) renewable resources should, within a specific area and time span, not exceed the formation of new stocks. Thus, for instance, yearly extraction of groundwater should not exceed the yearly addition to groundwater reserves coming from rain and surface water; and
- use of relatively rare nonrenewable resources, such as fossil carbon or rare metals, should be close to zero, unless future generations are compensated for current use by making available for future use an equivalent amount of renewable resources.

Georgescu-Roegen may be regarded as the first economist to formally advocate a link between entropy and economics. Most stocks of natural resources (e.g., crude oil) are found in states of low entropy and after utilisation in the production/consumption process, are released into the environment in the form of high entropy (e.g., CO_2). Georgescu-Roegen argues that production and consumption processes are time irreversible. According to him, 'the economic process is entropic: it neither creates nor consumes matter and energy, but only transforms low into high entropy' (Georgescu-Roegen, 1971:281). That is, the stocks of natural resources are permanently reduced or degraded by economic activities, and the stocks of waste products released into the atmosphere are permanently increased. Thus, in the absence of any intervention, it is implied that externalities will continue to increase as economic growth proceeds.

Georgescu-Roegen has extended the laws of thermodynamics by proposing a Fourth Law that considers the concept of **material entropy** in a closed system. **The 'Fourth Law' of Thermodynamics** states that in a closed system it is impossible to completely recover the matter involved in the production of work or wasted in friction. This implies that material entropy will always increase even if exergy (i.e., available energy) is plentiful. In the long run, material entropy will be maximised and will thus be unavailable for work. Georgescu-Roegen postulates that in the long-term there will be a '**material death**' of the economic system, which is similar to Clausius' heat death referred to earlier. The Fourth Law suggests that economic activities simply serve to increase entropy and that recycling is impossible. The Fourth Law implies, therefore, that economic growth is not sustainable because it degrades both energy and matter and leaves little available energy and matter for future generations.

Not everyone agrees with Georgescu-Roegen's views about economic growth and sustainability. In Chapter 11, we review both sides of this debate. Robert Ayres has taken issue with Georgescu-Roegen's Fourth Law of Thermodynamics. Ayres contends that the Fourth Law is not consistent with the laws of physics. He argues that given enough energy, any element can be recovered from any source and cites the recovery of gold from seawater as an example (Ayres, 1998). According to Ayres, 'in a closed system with a continuing supply of exergy, enough degraded (i.e., average) matter can be recycled and upgraded to maintain an effective materials extraction and supply system indefinitely' (Ayres, 1998:198).

In recent years, some economists have expressed caution about interpretation of the Second Law of Thermodynamics.[9] The point has been made that although non-renewable resources are finite in supply, they are not *the* constraint for the survival of humanity and the ecosystem. Many believe that technological progress could facilitate a shift from reliance on non-renewable to renewable energy. This would happen once we reach the point where the costs of extracting and refining natural resources exceed the cost of recycling.

Due to the fact that unavailable energy cannot be converted into exergy, exergy is a scarce factor and is more valued by economic agents. Some researchers have suggested that exergy should be used as an aggregate measure of environmental pollution. For example, Faber (1985), Kümmel (1989), and Ayres and Martinás (1995), have suggested that the exergy content of raw material inputs could be used as a measure of potential pollution from human economic activities. Göran Wall has used the exergy concept to measure the quality of all resources (renewable and non-renewable) in Sweden (Wall 1986).

Ayres (1998) has proposed a measure called **'exergetic efficiency'** which he defines as the ratio of exergy outputs to total exergy inputs including utilities. According to him, this measure could be used to provide an indication of the potential for future improvement of a process. Thus, for example, low exergetic efficiency would imply that process improvement could be used to reduce raw material and fuel inputs as well as waste products associated with the process. On the other hand, high exergetic

---

[9] See, for example, work by Ruth (1993, 1995); Biancardi *et al.* (1993a, 1993b); and Ayres, (1998).

efficiency would imply that the scope for future improvement is limited. Finally, Ayres suggests that exergetic efficiency could be used as a common measure (e.g., similar to say, GNP) that could be used for comparing different activities and processes. Other related measures such as **'heat equivalents'** (the amount of heat that is inevitably produced when cleaning the environment from the respective pollutant) and **'net energy'** have also been proposed.

Some ecologists (e.g., Ulanowicz and Hannon, 1987; Amir 1991) have gone a step further to suggest that entropy (or, rather exergy) should be used as a measure of the **value** associated with energy flows in the economy-environment system.[10] Under this proposal, the value of energy flow to the ecosystem would be calculated in terms of ecosystem prices and used to evaluate the efficiency of resource use. This idea has been roundly criticised by many economists, including Georgescu-Roegen himself. The main reason for their objection is that 'value', as used in economics, depends on the preferences of economic agents and not on the physical characteristics of material such as the level of entropy. To illustrate this important point, Georgescu-Roegen gives the example of a poisonous mushroom which has low entropy and yet is considered to be of value. To clarify his views on entropy and value, Georgescu-Roegen states that 'low entropy...is a necessary condition for a thing to have value. This condition, however, is not also sufficient' (Georgescu-Roegen, 1971:282).

2.5   Modelling Economy-Environment Interactions

A model can be defined as a simplified representation of reality. Parallels can be drawn between a model and a map. Like maps, models have different possible objectives and uses, and it is difficult to have a model that can serve a wide variety of uses (Robinson, 1991). Models are an abstraction from reality. In view of the fact that the real world is complex, abstraction is required in order to focus on the key variables. Thus, the level of complexity in a model is dependent on the purpose for which it is to be used. Costanza *et al.* (1995) have suggested that three criteria of models should be: realism, precision and generality. Realism is achieved by simulating the system behaviour in a qualitatively precise way, while precision is achieved by

---

[10] Odum (1971) and Costanza (1981) have also proposed the use of energy as the basis for market valuation.

doing so in a quantitatively precise way. Finally, generality is achieved by representing a broad range of systems' behaviours in the same model. However, in general, it is virtually impossible for a single model to capture all three criteria because there are trade-offs amongst them. For example, increasing precision will be at the expense of realism and vice versa.

Models can be classified according to their level of aggregation, starting from the individual or firm level, to the regional level and to the national or global levels. Economic models can also be classified into two broad categories: **partial equilibrium** and **general equilibrium** models. Partial equilibrium models consider the effect of changes in one explanatory variable on a single variable, holding all other variables constant. On the other hand, general equilibrium models consider the impact of changes in one (or many) variables on all other variables in the model simultaneously. For example, multiple regression models used to estimate supply and demand functions fall under the first category, whereas input-output (IO) models, computable general equilibrium (CGE) models, linear programming models, and system simulation models fall under the second category.[11]

In view of the fact that ecological economics involves the study of the interrelationships between the ecosystem and the economic system, the input-output framework proposed by Leontief (1941) was the popular model of choice in earlier economy-environment modelling work. In the IO approach, the environment is treated as one of the many sectors in the economy. Cumberland (1966) was among the first to apply the IO approach to the economy-environment system. He developed a model for estimating the cost of environmental utilization and purification. In subsequent studies, others (e.g., Ayres and Kneese, 1969; d'Arge and Kogiku 1973; Ayres and Noble, 1978) applied the mass balance principle within the IO framework.

Input-output models, based on the law of conservation of mass and energy, have been used to analyse the energy efficiency of production techniques, energy aspects of recycling and alternative sources of energy.[12] These models are useful for investigating the direct and indirect effects of material and energy substitution in the production processes. The IO model is consistent with the law of conservation of mass and energy because it requires inputs to be equal to outputs.

---

[11] General equilibrium and input-output models are referred to as 'economy-wide' models because they consider impacts on all sectors of the economy simultaneously.

[12] For example, see studies by Ayres (1989); Herendeen and Plant (1981) and Casler and Wilbur (1984).

Although the IO approach is useful for investigating aspects of the economy-environment interactions from the production side, it is not sophisticated enough to model economic behaviour. For example, it does not capture the preferences of economic agents which ultimately affect their consumption and production choices. Furthermore, the IO approach does not adequately account for the price system and changes in technology. Other restrictions of IO methodology are as follows:

- the input functions are linear and therefore do not allow for substitution between inputs;
- there is a one-to-one relationship between inputs and outputs, i.e., joint production is not allowed;
- in the static IO model, there are no capacity or capital restrictions. That is, the supply side of the economy is ignored.

Following the development of the IO model, there have been attempts to integrate both IO and econometric models into a single model. Such models are know as **Input-Output Econometric** (IOE) models (also referred to as **Micro-Macroeconometric Models** or **Integrated Models**. One of the earliest IOE models was constructed by Lawrence Klein (the 1980 Nobel laureate) in the early 1960s. The integration is achieved in the form of output from the IO component providing inputs to the econometric component. Current IOE models range in scope from cities to states (or regions) to countries, and to the global economy.

The deficiencies of the IO approach can be mitigated to some extent by specifying economy-environment interactions within a CGE framework. A recent trend in CGE modelling of economy-environment systems is to account for environmental aspects in the following two ways: (i) by modifying the model equations to account for the environmental effects; or (ii) by developing an environmental module to link to the economic module. An example of the former approach is work done on Sri Lanka (Bandara and Coxhead, 1999), the Philippines (Coxhead and Shively, 1995), and Indonesia (Strutt, 1998). An example of the latter approach is work done by Wiig *et al.* (2001) on Tanzania. In this study (see Box 2.1), a model of the nitrogen cycle in the soil was incorporated in a CGE model of the Tanzanian economy.

The major obstacles to developing economy-environment models include data acquisition and computational effort. The latter problem has

---

Box 2.1 An example of an economy-environment model

In this study, a model of the nitrogen cycle was linked to a CGE model of the Tanzanian economy, thus establishing a two-way link between the environment and the economy. The model consisted of 342 equations, of which 66 constituted the soil model and 276 described economic features of the CGE model. The economic model covered 20 production sectors, of which 11 were agricultural sectors. The equations in the soil model described the soil degradation process which is caused by soil mining and soil erosion. For a given level of natural soil productivity, it was assumed that profit-maximising farmers choose inputs (and hence production volumes) which in turn influence soil productivity in the following years through the recycling of nitrogen from the residues of roots and stover and the degree of erosion. The model was used to simulate the effects of typical structural adjustment policies such as a reduction in agro-chemical subsidies, reduced implicit export tax, and so on. Projections from the model indicated that after 10 years, Gross Domestic Product falls by more than 5% in the model with endogenous soil degradation, compared with a traditional CGE model with constant soil productivity.

Source: Wiig *et al.* (2001).

---

somewhat been lessened with the recent decline in the cost of computing power. However, the data requirement problem remains formidable. In order to link the environment to the economic system, one must be able to place a value on the goods and services provided by the ecosystem. There is an ethical problem here in the sense that some people think it is improper, or even immoral, to place a monetary value on things such as human life and biodiversity. However, many economists will argue that we need a common basis for comparing ecosystem goods and services with economic system goods and services. Also, as indicated earlier, most environment goods and services are often mistaken to be 'free' because they do not command a market price. 'Valuing' the environment enables us to more accurately evaluate competing alternatives.

The process of valuing ecosystem goods and services is in itself a controversial issue. We have already indicated above that ecologists' view of the value of the ecosystem, which is based on the laws of thermodynamics, is different from the economists' view, which is based on individuals' preferences. Costanza (1991) has suggested that what is needed is a pluralistic approach to valuation because a single approach yields imperfect information. In Chapter 5, we discuss the various types of values associated with the environment and recent methods that have been developed to estimate such values.

Norgaard (1989) has proposed the idea of an 'integrated, multi-scale, transdisciplinary, and pluralistic' (IMTP) approach to economy-environment modelling. The objectives of IMTP include the following:

- predicting the impacts of human activities on ecosystems;
- assessing the economic dependence on natural ecosystem services and capital, and
- modelling the integrated interdependence between ecological and economic components of the system.

IMTP models would allow policymakers to evaluate the temporal and spatial effects of, say, regional and global ecosystem response to regional and global climatic changes, acid rain precipitation and other environmental impacts. As already indicated, the data requirements for this type of modelling are massive. However, technological advances such as remote sensing and geographic information systems (GIS), as well as rapid developments in computing power and speed, have made it relatively easier to develop such models. What is required to make this happen is interdisciplinary co-operation and research funding.

2.6   Summary

In this chapter ecological economics has been defined as a sub-discipline of environmental economics that deals with the interrelationships between the economic system and the ecosystem. In addition to the emphasis on ecological processes, ecological economics considers social, ethical and political issues associated with resource use. The traditional (neoclassical) economic model was presented as a closed system that fails to properly account for flows from the environment. The main reason why the flow of goods and services provided by the environment tends to be ignored in economic models is that most environmental inputs are not traded in markets and therefore are often underpriced or unpriced. Traditional economic models also ignore the evolutionary and feedback processes associated with ecosystems.

An economy-environment system that accounts for flows of energy and materials was presented as a more realistic representation of economy-environment interactions. The environment in this system plays three important roles: as a provider of raw materials, as a receptacle for waste products and as a provider of amenities. The ability of the environment to assimilate waste products and provide additional services is severely restricted by indiscriminate creation of pollution.

The two major laws of thermodynamics were introduced in the chapter. The First Law of Thermodynamics, or the Law of Conservation of Mass and Energy, states that energy cannot be created or destroyed. The Second Law of Thermodynamics, or the Entropy Law, states that in a closed system entropy always increases. The concept of exergy, the maximum potential work that can be obtained from a system, was also introduced. The two laws were redefined in terms of exergy. In terms of the First Law, it can be said that the sum of exergy and unavailable energy remains constant. For the Second Law, it can be said that the exergy of an isolated system increases over time.

After defining the laws of thermodynamics, we considered their implications for the economy-environment system. The First Law implies that energy inputs equal energy outputs and that any raw material used for economic activities must eventually re-enter the ecosystem as waste products. Recycling cannot fully recover waste products. The Second Law implies that economic processes (i.e., production and consumption) are time irreversible in the sense that waste material cannot be fully converted to work. Because exergy is a scarce factor, some ecologists have suggested that it should form the basis of 'value' in economics. However, the concept of 'value' in economics is dependent on the preferences of individuals.

The IO approach (i.e., inputs equals outputs) is widely used in economy-environment modelling because it is consistent with the law of conservation of mass and energy. The computable general equilibrium approach was suggested as an improvement over the IO approach because it enables the economic sector to be modelled in a more realistic manner. The main difficulties in economy-environment modelling are data availability and computing resources to solve complex models. However, technological improvements such has GIS and satellite systems, as well as advancements in non-market valuation have made it more feasible to model economy-environment interactions.

In conclusion, the following points need to be stressed. Traditional economic models tend to ignore environmental effects. The laws of thermodynamics used within the context of economic models allow environmental effects to be accounted for. However, the laws of thermodynamics must be used with caution. For example, entropy by itself is incapable of fully explaining economy-environment interactions. There is therefore a need for a multidisciplinary approach in efforts to develop realistic models to assist sustainable development planning.

Key Terms and Concepts

autotroph	heterotroph
closed system	input-output econometric model
computable general equilibrium	input-output model
ecological economics	isolated system
economic system	law of conservation of mass and
economy-environment system	energy
entropy law	material death
evolution	materials balance
exergetic efficiency	micro-macroeconometric model
exergy	natural selection
first arrow of time	net energy
first law of thermodynamics	open system
general equilibrium	second arrow of time
heat equivalent	second law of thermodynamics

Review Questions

1. In your own words explain the meaning of the terms 'economic system' and 'ecosystem'.

2. Explain the difference between 'low-entropy energy' and 'high-entropy energy'.

3. Explain what is meant by isolated, closed, and open systems.

4. Explain the First and Second Laws of Thermodynamics.

5. Explain what is meant by the First and Second Arrows of Time.

Exercises

1. Explain the meaning of 'feedback' processes in an ecosystem. Give an example of positive and negative feedback within an ecosystem.

2. Select a large economy-environment system in your area. The ecosystem could be a bay, river, lake or forest. Construct a chart of material flows within this economy-environment system.

3. Read and summarise the views expressed by Ulanowicz and Hannon (1987) and Georgescu-Roegen (1971) concerning the use of entropy as a measure of value in the economy-environment system.

References

Amir, S. (1991). *Economics and Thermodynamics: An Exposition and Its Implications for Environmental Economics*. Resources for the Future, Quality of the Environment Division, Discussion Paper, No. QE92-04, Washington, D.C.

Ayres, R.U. (1998). Eco-thermodynamics: Economics and the Second Law. *Ecological Economics,* 26: 189-209.

Ayres, R.U. and Kneese, A.V. (1969). Production, Consumption, and Externalities. *American Economic Review,* 59: 282-297.

Ayres, R.U. (1989). *Energy Efficiency in the US Economy: A New Case for Conservation.* International Institute for Applied Systems Analysis, Laxenburg, Austria.

Ayres, R.U. and Martinás, K. (1995). Waste Potential Entropy: The Ultimate Ecotoxic. *Economie Appliquee,* XLVIII (2): 95-120.

Ayres, R.U. and Noble, S.B. (1978). Materials/Energy Accounting and Forecasting Models. In R.U. Ayres (ed.), *Resources, Environments, and Economics— Applications of the Materials/Energy Balance Principle.* John Wiley, New York.

Bandara, J.S. and Coxhead, I. (1999). Can Trade Liberalization Have Environmental Benefits in Developing Country Agriculture? A Sri Lankan Case Study. *Journal of Policy Modeling,* 21(3):349-374.

Barbier, E.B. and Rauscher, M. (1994). Trade, Tropical Deforestation and Policy Interventions. *Environmental and Resource Economics,* 4: 75-94.

Bergman, I. (1991). General Equilibrium Effects of Environmental Policy: A CGE Approach. *Environment and Resource Economics,* 1: 43-61.

Bianciardi, C., Tiezzi, E., and Ulgati, S. (1993a). Complete Recycling of Matter in the Framework of Physics, Biology and Ecological Economics. *Ecological Economics,* 8: 1-5.

Bianciardi, C., Tiezzi, E., and Ulgati, S. (1993b). On the Relationship Between the Economic Process, the Carnot Cycle and the Entropy Law. *Ecological Economics,* 8: 7-10.

Boulding, K.E. (1966). The Economics of the Coming Spaceship Earth. In *Environmental Quality in a Growing Economy: Essays from the Sixth RFF Forum,* ed, H. Jarret, John Hopkins Press, Baltimore, Md., pp.3-15.

Boulding, K.E. (1981). *Evolutionary Economics.* Sage, London.

Brooks, D.R. and Wiley, E.O. (1988). *Evolution as Entropy: Toward a Unified Theory of Biology.* The University of Chicago Press, Chicago, London, Second Edition.

Casler, S. and Wilbur, S. (1984). Energy Input-Output Analysis: A Simple Guide. *Resources and Energy,* 6: 187-201.

Clark, N. and Juma, C. (1987*). Long-Run Economics: An Evolutionary Approach to Economic Growth.* Pinter, London.

Costanza, R. (1981). Embodied Energy, Energy Analysis, and Economics. In H.E. Daley (ed.), *Energy, Economics, and the Environment.* Westview Press, Boulder, Colorado.

Costanza, R. (1991). Assuring Sustainability of Ecological Systems. In Robert Costanza (ed.), *Ecological Economics: The Science and Management of Sustainability.* Columbia University Press, New York.

Costanza, R, Daly, H.E. and Bartholomew, J.A. (1991). Goals, Agenda and Policy Recommendations for Ecological Economics. In R. Costanza (ed.), *Ecological Economics: The Science and Management of Sustainability.* Columbia University Press, New York.

Costanza, R., Wainger, L. and Bockstael, N. (1995). Integrated Ecological Economic Systems and Modelling: Theoretical Issues and Practical Applications. In J.W. Milon and J.F. Shogren (eds.), *Integrating Economic and Ecological Indicators.* Praeger, Wesport, Connecticut.

Coxhead, I. And Shively, G. (1995). Measuring the Environmental Impacts of Economic Change: The Case of Land Degradation in Philippine Agriculture. University of Wisconsin, Department of Agricultural Economics, Staff Paper Series No. 384, Madison, Wisconsin.

Crocker, T.D. (1995). Ecosystem Functions, Economics and the Ability to Function. In J.W. Milon and J.F. Shogren (eds.), *Integrating Economic and Ecological Indicators: Practical Methods for Environmental Policy Analysis.* Praeger, Westport, Connecticut.

Crocker, T.D. and Tschirhart, J. (1992). Ecosystems, Externalities and Economics. *Environmental and Resource Economics,* 2: 551-567.

Cumberland, J.H. (1966). A Regional Interindustry Model for the Analysis of Development Objectives. *The Regional Science Association Papers,* 17: 65-94.

Daly, H.E. (1991). *Steady-State Economics.* Island Press, Washington.

d'Arge, R.C. and Kogiku, K.C. (1973). Economic Growth and the Environment. *Review of Economic Studies,* 59: 61-77.

Dosi, G. and Nelson, R.R. (1994). An Introduction to Evolutionary Theories in Economics. *Journal of Evolutionary Economics,* 4: 153-172.

Faber, M. (1985). A Biophysical Approach to the Economy: Entropy, Environment and Resources. In W. van Gool and J. Bruggink (eds.), *Energy and Time in Economics and Physical Sciences.* Elsevier Science Publishers, North-Holland, Amsterdam.

Faber, M., Manstetten, R., and Proops, J. (1996) *Ecological Economics: Concepts and Methods.* Edward Elgar, Cheltenham, U.K.

Georgescu-Roegen, N. (1971). *The Entropy Law and the Economic Process.* Harvard University Press, Cambridge, Massachussetts.

Hannon, B. (1986). Ecosystem Control Theory. *Journal of Theoretical Biology,* 212: 417-437.

Hannon, B. (1991). Accounting in Ecological Systems. In Robert Costanza (ed.), *Ecological Economics: The Science and Management of Sustainability.* Columbia University Press, New York.

Herendeen, R. and Plant, R. (1981). Energy Analysis of Four Geothermal Technologies, and Labour Intensities for 1972. *Energy,* 6: 73-82.

James, D.E. (1993). Environmental Economic Industrial Process Models and Regional Residuals Management. In A.V. Kneese and J.L. Sweeney (eds.), *Handbook of Natural Resource and Energy Economics.* Vol. 1-3, North-Holland, Amsterdam.

Kneese, A.V., Ayres, R.U. and d'Arge, R.C. (1970). *Economics and the Environment.* Johns Hopkins University Press, Baltimore.

Kümmel, R. (1989). Energy as a Factor of Production and Entropy as a Pollution Indicator in Macroeconomic Modelling. *Ecological Economics,* 1: 161-180.

Layzer, D. (1976). The Arrow of Time. *Astrophysical Journal,* 206: 559-564.

Leontieff, W. (1941). *The Structure of the American Economy, 1919-1939.* Oxford University Press, Oxford.

Marshall, A. (1930). *Principles of Economics.* Macmillan, London.

Mayr, E. (1942). *Systematics and the Origin of Species*. Columbia University Press, New York.

Nordhaus, W.D. (1990). To Slow or Not to Slow: The Economics of the Greenhouse Effect. *Economic Journal,* 101: 920-937.

Nordhaus, W.D. (1993). How Much Should We Invest in Preserving Our Current Climate. In H. Giersch (ed.), *Economic Progress and Environmental Concerns*. Springer-Verlag, Berlin.

Norgaad, R.B. (1989). The Case of Methodological Pluralism. *Ecological Economics*, 1: 37-57.

Norgaard, R.B. (1984). Coevolutionary Development Potential. *Land Economics*, 60: 160-173.

Odum, H. (1971). *Environment, Power, and Society*. Wiley Interscience, New York.

Prigogine, I. and Stengers, I. (1984). *Order Out of Chaos*. Heineman, London.

Robinson, J.B. (1991). Modelling the Interactions Between Human and Natural Systems. *International Social Science Journal*, 130: 629-647.

Ruth, M. (1993). *Integrating Economics, Ecology and Thermodynamics*. Kluwer Academic Publishers, Dordrecht.

Ruth, M. (1995) Thermodynamic Implications of Natural Resource Extraction and Technical Change in US Copper Mining. *Environment and Resource Economics*, 6: 187-206.

Schrödinger, E. (1944). *What is Life?* Cambridge University Press, Cambridge.

Strutt, A. (1998). Trade Liberalisation and Land Degradation in Indonesia. ACIAR Indonesia Research Project, Working Paper 98.06, Centre for International Economic Studies, University of Adelaide.

Toman, Pezzey, J., and Krautkramer, J. (1994). Neoclassical Economic Growth Theory and Sustainability. In D. Bromley (ed.), *Handbook of Environmental Economics*. Blackwell, Oxford.

Ulanowicz, R.E. and Hannon, B. (1987). Life and the Production of Entropy. *Proceedings of the Royal Society of London,* B, 232: 181-192.

van den Berg, J.C.J.M. (1993). A Framework for Modelling Economy-Environment Relationships Based on Dynamic Carrying Capacity and Sustainable Development Feedback. *Environment and Resource Economics,* 3: 395-412.

van den Berg, J.C.J.M. and Nijkamp, P. (eds.), (1994). Sustainability, Resources and Region. *The Annals of Regional Science.* 28, special issue.

Wall, G. (1986) Exergy Conversion in the Swedish Society. *Energy,* 11: 435-444.

Wiig, H., Aune, J.B., Glomsrod, S. and Iversen, V. (2001). Structural Adjustment and Soil Degradation in Tanzania: A CGE Model Approach with Endogenous Soil Productivity. *Agricultural Economics*, 24: 263-287.

3.   How Markets are Supposed to Work

Objectives

After studying this chapter you should be in a position to:

❑   explain the characteristics of a competitive market;

❑   define a demand curve;

❑   explain the use of willingness-to-pay as a measure of benefit;

❑   define a supply curve and how it is derived from marginal cost;

❑   explain how equilibrium is achieved in a competitive market; and

❑   explain the concepts of consumer surplus and producer surplus

3.1   Introduction

Markets play an important role in our individual lives as well as in national, regional and global economies. In Chapter 2, we discussed how the activities of economic agents contribute to the generation of pollution. The operation of the market system is intimately related to the nature and amount of pollution generated. The purpose of this chapter is to introduce readers with a limited background in economics to the basic concepts of economic analysis. The intention is to give the reader an understanding of how markets work. This introduction is necessary in order to have a better appreciation of why markets tend not to work well for many environmental goods and services. The knowledge acquired would be useful in formulating solutions to environmental problems.

The following section describes the characteristics of a perfectly competitive market and the underlying assumptions. A brief discussion of consumer behaviour and the derivation of the demand curve follows. After that we take a look at the production side and how the supply curve is derived. Next, we bring together the demand and supply curves to explain how they interact to establish market equilibrium. Finally, we consider practical applications of the theory, including the notions of consumer and producer surplus.

3.2   The Competitive Market

A market can be defined as the coming together of consumers (or buyers) and producers (or sellers) to exchange goods and services for money.[13] In this technological age, the buyers and sellers do not have to be physically present to carry out transactions. Usually a market exists for a single good or service. However, it is common to find markets in which goods with similar characteristics are being offered for sale. Markets may be classified according to the numbers of sellers and buyers. In a **perfectly competitive** market, there are many buyers and sellers. A **monopoly** is one in which there is a single seller. An intermediate case is an **oligopoly** in which there are a few sellers. **Monopolistic competition** is a type of market structure in which there are many firms selling products that are close substitutes. Finally, a **monopsony** is a market in which there is a single buyer.

An example of a monopoly is the utilities sector in many countries where the government is often the sole provider of water and electricity. The Australian domestic car manufacturing market is an example of oligopolistic competition because there are four main sellers: GM Holden, Ford, Toyota and Mitsubishi. An example of monopolistic competition is the fashion industry where there are many firms selling similar items of clothing that can only be differentiated by brand name. An example of a monopsony is that of a small town where there is only a single major industry, say, a mine. In this case, the mine is the sole buyer of labour and other goods and services.

The competitive market has the following characteristics:

---

[13] There could also be third parties such as brokers and agents who are considered as an essential part of a market.

- there are many buyers and sellers and none of them are influential enough to affect the market price or output;
- the buyers and sellers are free to enter and leave the market in response to price changes;
- the goods and services being offered for sale are identical (i.e., homogeneous). This implies that buyers do not care from whom they buy, provided prices are identical; and
- all the participants in the market have perfect knowledge. That is, consumers know product prices and producers know input prices.

3.3　Consumer Behaviour and Demand

In this section, we consider the behaviour of an individual consumer which gives rise to an individual demand curve. We discuss how the demand curve can be used to value benefits. Finally, we consider the case of market demand which is essentially an aggregation of market demand.

3.3.1　The Individual Demand Curve

The term 'consumer' is used here to describe a typical person who has to decide what combinations of goods, services and amenities to purchase, given a fixed budget of money income and time. In this chapter, we will be concerned with **private goods.** These are goods that are **rival** in consumption. That is, one person's consumption precludes consumption by another. In Chapter 4, we will consider **public goods**, where one person's consumption does not does not diminish another's ability to consume.

We will assume that consumers derive 'utility' or satisfaction from consuming goods and services. We will consider a 'benefit' to be the value or gain in utility that a consumer obtains from consuming a good or service. Let us presume that the consumer's utility function can be represented by the following equation:

$$U = f(q_1, q_2, ...q_n) \tag{3.1}$$

where $q_1, ... q_n$ are quantities of the goods, services and amenities that yield satisfaction to the consumer. This function will be unique for each individual. We make further simplifying assumptions about the consumer.

1. The consumer has full information about alternative bundles of goods, services and amenities he can choose from.
2. The consumer has the ability to consider all the alternatives and develop a consistent preference ranking among them. For example, if she prefers good A to good B, and prefers good B to good C, then, to be consistent, she must prefer A to C.
3. The consumer's preference ranking for a good is stable, at least in the period of the analysis; and
4. The consumer's preferences are consistent with the relative amounts of utility that each alternative yields.

The last assumption is necessary to ensure that the consumer's preference relationships referred to in (2) are also consistent with her utility rankings.[14] For example, in (2), it can be inferred that A's utility will be greater than B's utility; B's will be greater than C's, and A's will also be greater than C's.

We assume that the consumer chooses a combination of goods and services so as to maximise total utility subject to an income constraint.[15] The result of this choice problem is a **demand function,** which is essentially a schedule or a curve that indicates how much of a good a consumer will buy at various prices. For example, Table 3.1 presents a demand schedule for doughnuts. Let individual A's demand for doughnuts be expressed by the function, $q_1^D = -2p + 10$. Using this function, we can determine how much individual A will buy at various prices. For example, if doughnuts are free, A will demand 10 doughnuts. However if the doughnuts cost \$5 or more each, she will not buy any. To plot A's demand curve, we use the inverse form of the demand function, that is, $p = -0.5q_1^D + 5$. This function generates the curve in Figure 3.1.

---

[14] Some of these assumptions are quite strong and others may be unrealistic in some situations. However, it is an example of modelling whereby we simplify reality in order to account for complex phenomena.

[15] We return to this choice problem in Chapter 6.

Table 3.1 A hypothetical demand schedule for doughnuts

Price per doughnut ($)	Quantity demanded $(q^D_1 = -2p + 10)$
0.00	10
0.50	9
1.00	8
1.50	7
2.00	6
2.50	5
3.00	4
3.50	3
4.00	2
4.50	1
5.00	0

Figure 3.1 A hypothetical individual demand curve for doughnuts

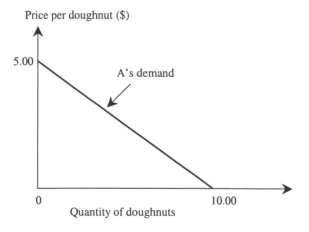

Note the **inverse relationship** between price and quantity demanded. This is referred to as the **Law of Demand**. That is, given income, preferences and prices of alternative goods, an individual will be willing to purchase decreasing amounts of a given good (or service) as its price increases.

In addition to the downward slope, there are two other points to note about the demand curve:

1. The individual's demand for good q_1 is defined given that all other goods, in this case, q_2, and income remain constant.
2. The demand curve is defined for a given period of time. Thus, the demand curve in a different period of time will have a different shape and position.

We began this section by saying that a benefit is the value or gain in utility the consumer obtains from consuming a good or service. For this reason, points on the demand curve represent the maximum amount of money the consumer is willing to pay for different quantities of the good. **Willingness-to-pay** (WTP) is a measure of satisfaction or utility. In Figure 3.1, as the individual consumes more doughnuts, her WTP for an extra doughnut declines. The WTP curve also defines the benefits to society from consuming the given good or service. We return to this issue later in the chapter.

3.3.2 The Concept of Elasticity

The term **'elasticity'** refers to the responsiveness of quantity demanded (or supplied) to changes in other variables (e.g., price and income). The concept of elasticity is important because a key factor in the functioning of the economic system is the reaction of economic agents to price incentives. The extent to which consumers (producers) react to price signals depends on the magnitudes of the elasticities of demand (supply).

Own-price elasticity of demand is the ratio of the change in quantity demanded of a given good to the change in its own price. That is,

$$\varepsilon^D = \frac{\%\ \text{change in quantity of } q_1 \text{ demanded}}{\%\ \text{change in price of } q_1} = \frac{\Delta q_1 / q_1}{\Delta p_1 / p_1} \qquad (3.2)$$

Depending on the magnitude of the elasticity parameter, own-price elasticity of demand can be:

*   perfectly elastic
*   relatively elastic
*   relatively inelastic or
*   perfectly inelastic

1. If $|\varepsilon^D|=\infty$, demand is perfectly elastic. The demand curve, in this case, is horizontal (Figure 3.2, Panel a). A small increase in the price of the good will cause quantity demanded to fall to zero. In practice, no good has perfect price elasticity.
2. If $|\varepsilon^D|>1$, demand is relatively elastic giving rise to a relatively flat demand curve (Figure 3.2, Panel b). In this case, a small change in the price of the good causes a relatively large change in quantity demanded. In general, most luxury goods tend to be relatively price elastic.
3. If $|\varepsilon^D|<1$, demand is relatively inelastic. The demand curve in this case is relatively steep (Figure 3.2, Panel c). In this case, a change in the price of the good causes little change in quantity demanded. Necessities such as food and utilities (e.g., water and energy) tend to be relatively price inelastic.
4. If $|\varepsilon^D|=0$, demand is perfectly inelastic, the demand curve is thus vertical (Figure 3.2, Panel d). A change in the price of the good does not lead to a change in quantity demanded.

Cross-price elasticity of demand measures the responsiveness of the quantity of q_1 demanded as a result of changes in the price of another good, for example, q_2. Cross-price elasticity of demand is given by:

$$\varepsilon^D_{12} = \frac{\% \text{ change in quantity of } q_1 \text{ demanded}}{\% \text{ change in price of } p_2} = \frac{\Delta q_1 / q_1}{\Delta p_2 / p_2} \tag{3.3}$$

The magnitude of cross-price elasticity allows us to classify goods as **substitutes** or **complements**. For example,

1. $\varepsilon^D_{12} > 0$, implies q_1 and q_2 are substitutes. That is, an increase in the price of one good causes consumers to switch to the other, resulting in an increase in the quantity demanded of the second good. Examples of substitute goods are sugar and nutrasweet, bus and rail transportation, and so on.
2. $\varepsilon^D_{12} < 0$, implies q_1 and q_2 are complements. Complementary goods are consumed together and therefore an increase in the price of one good leads to a reduction in its consumption, and hence a reduction in demand

for the other good. Examples of complementary goods include bread and margarine, beer and nuts, and so on.

Figure 3.2 Elasticity of demand

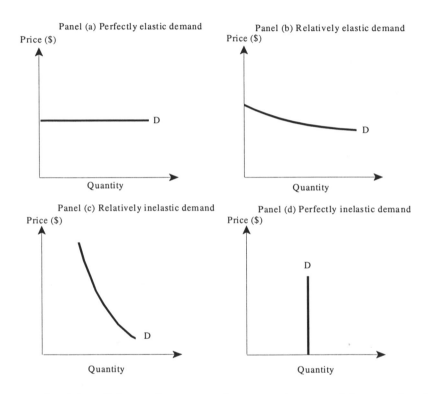

Income elasticity of demand measures the responsiveness of the quantity of a good demanded given a change in income. It can be defined as:

$$\eta_Y = \frac{\% \text{ change in quantity of } q_1 \text{demanded}}{\% \text{ change in income}} = \frac{\Delta q_1 / q_1}{\Delta Y / Y} \qquad (3.4)$$

1. $\eta_Y > 0$, implies the good is a **normal good**. Most goods are normal goods because an increase in income leads to an increase in quantity demanded.

2. $\eta_Y < 0$, implies the good is an **inferior good**. That is, an increase in Y leads to a decrease in q_1. There are not many practical examples of inferior goods. However, a low-income family that currently consumes dried vegetables might reduce their consumption and switch to fresh vegetables in response to an increase in household income.

3.3.3 Market (Industry) Demand

The market demand curve for private goods is obtained by a horizontal summation of the individual demand curves for the particular product. For example, suppose that the market for doughnuts comprises two individuals, A and B. We have already defined A's demand function. Let B's demand function be defined by $q^D_2 = -2p+12$. Market demand is obtained by adding up the quantities demanded by A and B at each price. This is referred to as **horizontal summation**. For example, when $p > \$6.00$, market demand will be zero; at $\$5 < p \le \$6$, A's demand will be zero and market demand will be given by B's demand which is:

$$q^D = -2p + 12$$

At $p \le \$5$, market demand function is given by:

A's demand:	$q^D_1 = -2p + 10$
+ B's demand:	$q^D_2 = -2p + 12$
= Market demand:	$q^D = -4p + 22$

The quantities demanded at various prices are shown in Table 3.2. It can be seen that if doughnuts are free, A will demand 10 and B will demand 12, resulting in a total demand of 22 doughnuts. However, at a price of $6.00 each, none will be demanded.

At a price of $5.00, A will purchase none, whereas B will purchase 2 doughnuts. Market demand is therefore 2 doughnuts. This exercise is repeated for every other price to obtain the remaining points on the market demand curve. Figure 3.3 presents the inverse market demand curve generated from Table 3.2.

Table 3.2 A hypothetical market demand schedule for doughnuts

Price per doughnut ($)	A's quantity demanded ($q^D_1 = -2p + 10$)	B's quantity demanded ($q^D_2 = -2p + 12$)	Market demand (A + B) ($q^D = -4p + 22$)
0.00	10	12	22
0.50	9	11	20
1.00	8	10	18
1.50	7	9	16
2.00	6	8	14
2.50	5	7	12
3.00	4	6	10
3.50	3	5	8
4.00	2	4	6
4.50	1	3	4
5.00	0	2	2
5.50	0	1	1
6.00	0	0	0

Figure 3.3 A hypothetical market demand curve for doughnuts

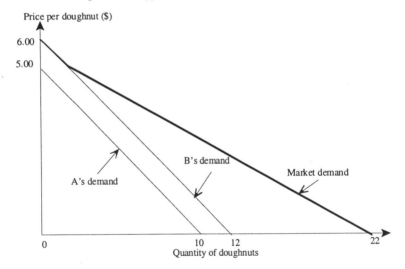

3.4 Producer Behaviour and Supply

Suppose a producer uses inputs (raw materials and other goods) to produce one good or service for sale. We can express the production by means of a mathematical equation or **production function**. That is:

$$q = f(x_1, x_2, \ldots, x_n) \qquad (3.5)$$

where $x_1, x_2, \ldots, x_n$ are various inputs (e.g., labour, land and capital) used in producing the good, q. We will assume that the producer's aim is to maximise profit subject to the constraint of the above production function. Given the profit motive, the producer will increase the output of q if its price rises so as to increase profits. A supply schedule indicates how much of a good a producer will supply at various prices. Let the supply function for doughnuts be given by the equation, $q^S = 2p - 2$. Table 3.3 shows how many doughnuts will be supplied at various prices.

Table 3.3 A hypothetical supply schedule for doughnuts

Price per doughnut ($)	Quantity supplied ($q^S = 2p - 2$)
1.00	0
1.50	1
2.00	2
2.50	3
3.00	4
3.50	5
4.00	6
4.50	7
5.00	8
5.50	9
6.00	10

Figure 3.4 shows the supply curve generated from the data in Table 3.3. The supply curve may refer to an individual producer or all producers in the industry. Let us assume that this particular curve is the industry supply curve. We can make the following observations about it:

1. It is upward sloping (i.e., positively sloped) because producers are willing to supply more as price increases.
2. The supply curve refers to a given point in time. In this example, the supply curve for doughnuts for a different period may have a slightly different shape and position.

Figure 3.4 A hypothetical supply curve for doughnuts

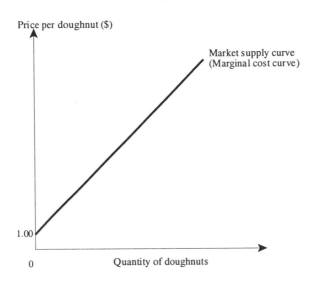

The supply curve is also the **marginal cost** (MC) curve. That is, it indicates the cost of producing each additional unit of the good. In order to maximise profits, the producer will increase production up to the point where **marginal revenue** (MR), the price per unit of output in a competitive market, just equals marginal cost.

3.5   Market Equilibrium in the Competitive Market

In this section, we bring together the demand and supply curves to explain how market equilibrium is attained. The interaction of supply and demand forces in the market determines the **equilibrium** or **market clearing price**, and the **equilibrium quantity** demanded. The equilibrium price, in turn, determines the price for each unit of output, that is, marginal revenue. We assumed in Section 3.2 that in the perfectly competitive market, producers cannot affect the market price. Thus, in this case, the marginal revenue curve will be a horizontal line, which is the same as the price line. In Chapter 4, we show that this assumption is not valid in the case of a monopoly.

Table 3.4 puts together the market demand and supply schedules. Note that market equilibrium is achieved at a price of $4.00 at which market demand (6 doughnuts) is exactly equal to the quantity the market is willing to supply. To explain why $4.00 is the equilibrium price, suppose the price per doughnut is as high as $5.00. At this price, the producer will be willing to supply 8 doughnuts but consumers will demand only 2 doughnuts. Therefore there will be a demand deficit (or excess supply) of 6 doughnuts in the market. To clear the excess supply, the producer will be forced to mark down the price of doughnuts. As the price falls, consumers will increase their purchases, and producers will reduce supply. At $4.00 per doughnut, prices clear the market as quantity demanded will be equal to quantity supplied.

Table 3.4 Market demand and supply schedules for doughnuts

Price per doughnut ($)	Market demand $(q^D = -4 + 22)$	Market supply $(q^S = 2p - 2)$	Demand surplus (+) or deficit (−)
0.00	22	0	22
0.50	20	0	20
1.00	18	0	18
1.50	16	1	15
2.00	14	2	12
2.50	12	3	9
3.00	10	4	6
3.50	8	5	3
4.00	**6**	**6**	**0**
4.50	4	7	−3
5.00	2	8	−6
5.50	1	9	−8
6.00	0	10	−10

Now suppose the market price per doughnut falls to $1. At this price, consumers will demand 18 doughnuts but producers will not be willing to supply any doughnuts, resulting in surplus demand of 18 doughnuts. Given that the amount demanded is greater than the supply, there will be a shortage of doughnuts in the market. This will put upward pressure on prices and the producer will respond by putting more doughnuts on the market. As the price goes up, consumers will reduce their purchases until the price reaches $4.00, at which point, quantity demanded will be equal to quantity supplied.

Figure 3.5 shows a graphical representation of the market equilibrium. Note that equilbrium is at the point of intersection of the demand and supply curves. Factors that can shift the demand curve include income, prices of substitutes/complements and consumer tastes and preferences. An increase in income, with all other factors held constant, causes an upwards (i.e., rightward) shift in the demand curve for a normal good, whereas a decrease in income will shift it downward (i.e., leftward).

Figure 3.5 Equilibrium in the market for doughnuts

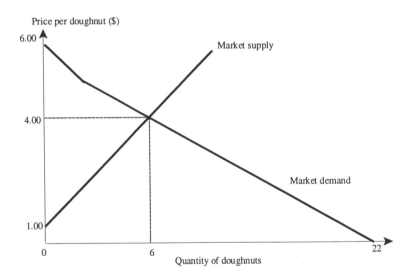

For example, an increase in per capita income in a given population (with all other factors constant) will shift the demand curve for mobile telephones upward (Figure 3.6, Panel a). However, due to the excess demand, the price of mobile telephones will rise to re-establish equilibrium.

A decrease in the price of substitutes for a good will cause the demand curve for the good to shift downward; a decrease in the price of complements for the good will cause the demand curve to shift upward. For example, a decrease in, say, the price of oil will cause a downward shift in the demand curve for natural gas (Figure 3.6, Panel b). The quantity of natural gas demanded falls and the price also falls to re-establish equilibrium.

Figure 3.6 Shifts in the demand curve

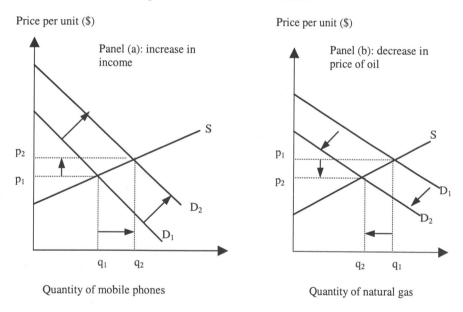

Price per unit ($)

Panel (a): increase in income

S

p_2

p_1

D_2

D_1

q_1 q_2

Quantity of mobile phones

Price per unit ($)

Panel (b): decrease in price of oil

S

p_1

p_2

D_1

D_2

q_2 q_1

Quantity of natural gas

Factors that can cause the supply curve to shift include price of inputs, taxes, subsidies, improvements in technology and weather (for agricultural products). A decrease in the price of inputs for making good q will cause the supply curve for q to shift outward, whereas an increase in the price of inputs will cause an inward shift. Improvements in technology will cause an outward shift because more output can be produced with the same level of inputs. For agricultural and other forms of production that are weather dependent, deterioration in weather conditions will cause a leftward shift in the supply curve. For example, a fall in the price of computer chips will shift the supply curve for computers outward (Figure 3.7, Panel a). Given the excess supply, the price for a computer will fall and the quantity demanded will increase.

Figure 3.7 Shifts in the supply curve

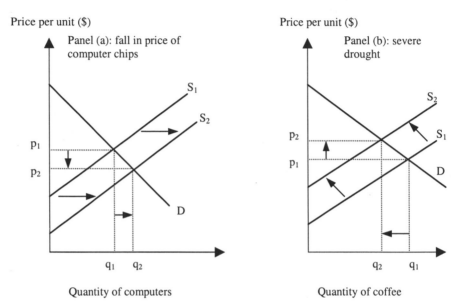

Another example is the world supply of coffee. A prolonged drought in a coffee growing area will shift the world supply curve for coffee inward. With the reduced supply of coffee, there will be excess demand and therefore the world price of coffee will rise to re-establish market equilibrium.

The example of equilibrium illustrated above is highly simplified in order to convey the basic concepts. In real life, equilibrium does not tend to be static. The demand curve is constantly shifting due to changes in tastes and incomes, while the supply curve also shifts in response to resource constraints and technological advances. Nevertheless, the analysis we have presented is useful in trying to predict the direction of change of prices and quantities. One important assumption underlying the market equilibrium analysis above is that property rights are well defined. That is, the seller owns the rights to the good or service and can therefore appropriate any benefits from the sale. We argue in the next chapter that this assumption breaks down in the case of environmental goods due to lack of well-defined ownership rights or the public goods nature of such goods.

3.6   Consumer and Producer Surplus

Consumer surplus is the maximum amount of money consumers are willing to pay for the good or service less the market price. To give an example of consumer surplus, let us revisit the doughnut example (see Figure 3.8). Consumer surplus is given by triangle *abc*, which is made up of two triangles and a rectangle whose areas are shown in the figure. The value of consumer surplus is therefore given by:

$$1 + 2 + 2 = \$5.00$$

This example shows that consumer surplus is a measure of net benefits or welfare and that reliance on only the market price could result in an under estimation of benefits.

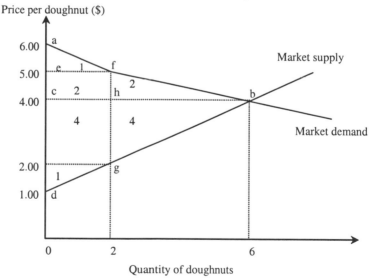

Figure 3.8 Consumer and producer surplus

Producer surplus is the net benefits received by the producer and is given by the difference between the market price and marginal cost, area *bcd*. The producer surplus is given by:

$$4 + 4 + 1 = \$9.00$$

Total benefits to the society (consumer plus producer surplus) are therefore given by area *abc* plus area *bcd*, which is equal to area *abd*. That is, $5.00 + $9.00 = $14.00. Now, suppose the government decides to set the price of a doughnut at $5.00. At this price, 2 doughnuts will be purchased and consumer surplus will be reduced to area *aef*, or $1.00. Producer surplus is now given by the area *defg*, or 2 + 4 + 1 = $7.00.

The net benefits of this policy are therefore given by sum of consumer plus producer surplus. That is, $1.00 + $7.00 = $8.00. Under this government policy, there is a net welfare loss of $14.00 − $8.00 = $6.00. This is referred to as a deadweight loss to society, and is defined by triangle *fbh*.

3.7   Applications of the Competitive Model

In this final section, we apply the basic economic principles to real-life economic problems. We consider two problems: (a) the socially optimal level of forestry; and (b) the effect of pollution abatement costs on employment.

3.7.1   The Socially Optimal Level of Forestry

Clear felling of timber has several undesirable effects on society. Examples include loss of forest cover and associated problems such as increased soil erosion, loss of soil nutrients, loss of biodiversity and so on. In Chapter 4, we attribute this problem to market failure. That is, the stumpage price (i.e., price charged per log) only considers the private marginal cost and does not include the external costs imposed on society. In this section, we analyse a policy of trying to address this problem by including environmental (or external) costs in the stumpage price.

Assume that the stumpage price is currently p_1. At this price, q_1 logs will be harvested. Now suppose that the government charges an extra $5 per log to cover the environment damage. This policy will result in an upward shift in the supply curve from S to S^* by a vertical distance of $5 (Figure 3.9). Assuming the demand for logs remains constant during the period of the analysis, the quantity of logs harvested will decline and equilibrium will be re-established at q^*, the socially 'optimal' level.[16]

Figure 3.9 Socially optimal level of forestry

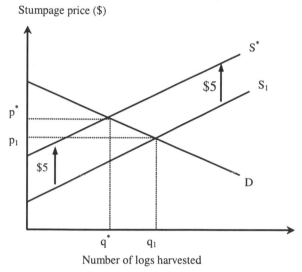

Number of logs harvested

3.7.2   The Effect of Pollution Abatement Costs on Employment

Suppose the supply curve for labour in an industry is S_L and the demand for labour is D_L. We consider the effect of increasing emission standards on employment under two scenarios: a flexible labour market and fixed wage (i.e., minimum wage) market. Let us first consider the situation where the government has set a minimum wage policy (Figure 3.10, Panel a). In this case the supply curve for labour is w_2aS. With the imposition of stricter environmental standards, the demand curve for labour shifts from D_L to D'_L and the equilibrium employment level falls from L_1 to L_2. Next, consider a labour market where wages are flexible. The equilibrium wage rate is initially w_1 and the equilibrium employment level is L_1 (Figure 3.10, Panel b). If the government raises the emission standard, some firms will be forced to leave the industry because they can no longer afford to cover their

---

[16] 'Optimal' is used here in an economic efficiency sense and does not necessarily mean this level is just or equitable.

variable costs. Consequently, the demand for labour will shift from D_L to D'_L. The downward shift in the demand curve will cause a surplus of workers looking for work at the old wage rate, w_1, resulting in a decline of the wage rate to w_2 and a new equilibrium employment of L_2.

Figure 3.10 The effect of emission standards on employment

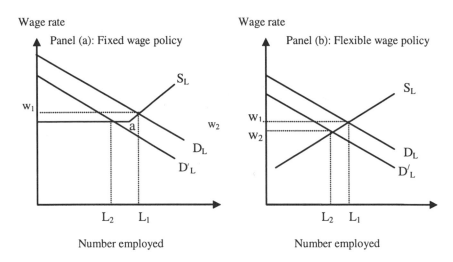

In terms of welfare effects, job losses under a minimum wage policy are greater than under a flexible wage policy. The two scenarios may also be compared in terms of the size of the consumer surplus. The consumer surplus in the case of minimum wages is smaller than in the flexible wage scenario.

3.8   Summary

In this chapter we have presented a model of how consumers and producers behave in a competitive market system. This system is characterised by many sellers and buyers such that no particular economic agent has influence in the market and everyone has perfect knowledge or information. Consumers seek to maximise their utility or satisfaction, whereas producers maximise profits. The outcome of the consumer's maximisation problem, given prices and income, is a demand function which is negatively sloped. The outcome of the producer's maximisation problem, given input prices and the production function, is the supply or marginal cost curve which is positively sloped. That is, the producer is willing to supply more given higher prices. Market equilibrium is achieved when quantity demanded exactly equals quantity offered for sale.

We explained that the demand curve is defined, in a given period, by income, prices and consumer tastes and preferences. A change in any of these factors will cause the demand curve to shift. The supply curve is also defined, in a given period of time, by input prices and technology. A change in any of these factors will cause the supply curve to shift. A change in product prices, will, *ceteris paribus*, cause a movement along the demand or supply curve. Finally, it was explained that the correct measure of social benefits is the sum of consumer and producer surplus. Consumer surplus is a measure of net benefits to the consumer and is defined by the difference between the maximum amount a person is willing to pay for a good over and above the market price. Producer surplus is a measure of net benefits (or profit) to the producer and is defined by the difference between the market price and marginal cost.

Key Terms and Concepts

consumer surplus	monopoly
cross-price elasticity of demand	monopsony
demand function	normal good
horizontal summation	oligopolistic competition
income elasticity of demand	oligopoly
individual demand	own-price elasticity of demand
inferior good	perfect competition
law of demand	private goods
marginal cost	producer surplus
marginal revenue	production function
market demand	supply elasticity
market equilibrium	supply function
monopolistic competition	willingness-to-pay

Review Questions

1. State the differences between perfect competition, monopoly, monopsony, oligopoly and monopolistic competition and give a practical example of each.

2. Explain the principle underlying the negative slope of a demand curve.

3. Explain the terms 'normal good' and 'inferior good'.

4. What is the meaning of the term 'elasticity' in economics?

5. Explain why the supply curve is also referred to as the marginal cost curve.

6. What factor (or factors) causes a movement along the demand curve and a demand shift?

Exercises

1. Indicate whether the demand for each of following goods is relatively price elastic or relatively price inelastic. State your reasons.

    (a) Electricity.
    (b) Mobile telephones.
    (c) Liquefied petroleum gas with respect to the price of oil.
    (d) Petrol.
    (e) Compact disc players.

2. Give a rough indication of the magnitude (i.e., whether greater than or less than 1.0) of the income elasticities of demand for the following commodities, stating your reasons.

    (a) A trip to Disneyland.
    (b) A trip to see the dentist.
    (c) Bread.

(d) Pizza.

3. Suppose the demand for and supply of water treatment equipment in a certain city can be represented by the following equations:

   Demand: $p = 180 - 4q^D$
   Supply: $p = 6q^S$

   (a) Determine the equilibrium price and quantity.
   (b) Describe what will happen in the market if the price of water treatment equipment is $40 per unit.
   (c) Suppose the government decides to impose a sales tax of $20 on each unit. Describe what will happen in the market and determine the new equilibrium quantity.
   (d) Suppose that before the sales tax the government decides to make the equipment affordable and therefore legislates a maximum price (price ceiling) of $48. Describe what will happen in the market.

4. Is it true that whenever a tax is introduced producers pass on all of the tax to consumers? State your reasons. Under what conditions would this be possible?

5. Explain why the market price of a good is not an accurate measure of the benefits derived from consuming a good.

6. Referring to Question 3, suppose the government imposes a carbon tax of 10%. Calculate the following:

   (a) Consumer surplus.
   (b) Producer surplus.
   (c) The net benefits of the policy.
   (d) Deadweight loss.

Further Reading

Frank, R.H. and Bernanke, B.S. (2004). *Principles of Microeconomics.* Second Edition, McGraw Hill, Irwin.

Mankiw, N. G. (1998). *Principles of Microeconomics*. Dryden Press, Fort Worth, Texas.

Pindyck, R.S. and Rubinfeld, D.L. (2001). *Microeconomics*. Fifth Edition, Prentice-Hall, Upper Saddle River, New Jersey.

4.   Why Markets 'Fail'

Objectives

After studying this chapter you should be in a position to:

❑   explain what is meant by property rights

❑   explain the meaning of an externality

❑   explain why market failure occurs

❑   explain why policy (government) failure occurs

❑   propose solutions to the problem of market failure

4.1   Introduction

In the previous chapter we explained how a market functions to allocate resources efficiently. A number of assumptions were made in order to obtain this result. It was implicitly assumed that the market price of a good (or service) reflects its **opportunity cost**. Opportunity cost is defined here as the value of a resource in its next best alternative use.[17] It was assumed that both consumers and producers have perfect information about prices and other relevant variables. It was also implicitly assumed that sellers in the market have well-defined ownership or property rights to the goods and services offered for sale. Given these conditions associated with a perfectly competitive market, the pursuit of self-interest by both consumers and

---

[17] Suppose a piece of forestland has two possible uses: harvesting the forest for its timber or leaving it as a natural environment. If the first use is chosen, the opportunity cost of the timber is the benefits that would have accrued if the forest had been left in its natural state.

producers results in an efficient allocation of resources. 'Efficiency', in this sense, is also referred to as **Pareto optimality**. That is, when the market reaches equilibrium, it is impossible to make anyone better-off without, at the same time, making at least one person worse off. In other words, at the equilibrium point it is impossible to reallocate or redistribute resources in more efficient ways. This chapter discusses the issues of market and policy failure, focussing on why they occur and what can be done about it. The chapter is organised into seven sections. Section 2 outlines the various types of market failure, followed by a brief discussion of each type in Sections 3 to 5. Section 6 conducts discussion of some solutions to address the problem of market failure, while the final section contains the summary.

4.2   Types of Market Failure

As indicated above, markets work well when prices reflect all values, that is, opportunity cost. 'Market failure' is said to occur when some costs and/or benefits are not fully reflected in market prices. The market system fails to function properly for many kinds of environmental goods because such resources, including the services they provide, are often not traded in markets. Thus, in general, market prices do not fully reflect the value of environmental goods and services. We demonstrate below that market failure results in an inefficient allocation of resources. Market failure can occur due to any or all of the following:

- lack of or weak property rights
- public goods and/or common property characteristics
- externalities, and
- type of market structure

4.2.1   Lack of or Weak Property Rights

Imagine a society where there is public ownership of all means of transport. An individual may drive any available car he or she sees on the roads, drive to their destination, and abandon the car by the roadside for use by other users. Under this scenario, people are likely to purchase just enough fuel to carry them to their next destination and only basic or inexpensive repair work, if any, would be carried out on a car. It is not difficult to envisage that after a while, there would be a lot of broken down cars by the roadside. This idealised example illustrates that when property rights are not well-defined or absent in the economic system, there is no incentive for a 'rational' individual to invest in an asset because they cannot appropriate the full benefits.[18]

Characteristics of property rights

When one purchases an asset such as a motor vehicle, it comes with a set of well-defined ownership rights and responsibilities. These ownership rights and responsibilities have the following general characteristics:

- **Well-defined:** You have title to the car in the form of a motor vehicle registration certificate and/or a purchase receipt. In some cases the entitlement may be informal and may have been institutionalised by social or cultural norms.
- **Exclusive:** You are the only one who has the right to use the car, although you may choose who else may use it and under what conditions. It is important to note that restrictions accompany ownership rights. You can drive on different sections of the roads at specified speed limits.
- **Transferable:** You may transfer permanent rights to the car to someone else by selling it, or you may transfer temporary rights by renting it. Note that the concept of completely transferable rights carries the notion of **'divisibility'**. That is, different types of rights associated with the asset may also be transferable. For example, your ownership rights to a block of land may be divided into parcels and each transferred. In this case you can choose to lease, rent or sell outright different parcels. It is also important to bear in mind that acquiring ownership rights to an asset involves a transfer of rights rather than a physical transfer of the asset. For example, when you buy a piece of land you do not take it home with you.
- **Secure and enforceable:** Your title to the car is secure and enforceable because if someone steals it, you have the right to

---

[18] The economic model assumes that each economic agent acts in his or her self-interest and that altruistic motives are absent. In Chapter 5, this assumption is relaxed to include situations where other motives underlie preferences.

notify the police and have the person arrested. Effective enforcement involves the following: effective detection and apprehension of violators and application of an appropriate penalty. To be effective, the penalty must exceed the actual or anticipated benefits of violation.

Property rights regimes

In practice, there are different types of property rights regimes. The complete set of property rights that we have described above is at one end of the spectrum and is typical of **private goods**. At the other end of the spectrum, there are **public goods**, with **congestion goods** in between these two extremes. Public goods can be classified into **pure public goods, semi-public goods**, and **open access** or **common property goods** (Figure 4.1). We define these different types of goods later.

Figure 4.1 A taxonomy of environmental goods

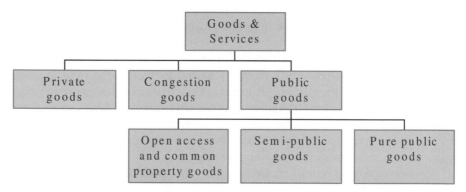

Most environmental goods fall under the category of pure public goods or open access/common property goods. In such cases lack of well-defined property rights results in market failure. A consequence of market failure is inefficient allocation of resources (e.g., excessive pollution). For example, a farmer has the right to prevent someone from polluting his or her land, but cannot prevent anyone from polluting a nearby river. Furthermore, he or she may have no legal right to receive compensation from the upstream polluters. The upstream polluters, who do not bear the costs of their activities, have no economic incentives to limit the amount of pollution especially when they know that the farmer has no property rights. This type of market failure has led to calls for governments to intervene in the market. We consider these solutions later in this chapter.

4.2.2 Public Goods

In this section we first define three broad categories of goods: private, congestion and public goods. We then go on to discuss the different types of public goods.

Private and pure public goods

As defined above, a private good has characteristics such as exclusivity, transferability, security and enforceability. In addition, a private good has a positive marginal cost. That is, the cost of supplying one additional unit is above zero. As was indicated in Chapter 3, a private good is **rival** in consumption. That is, once someone consumes the good, another person cannot consume it. On the other hand, a pure public good is **non-exclusive** and **non-rival** in consumption, and has **zero marginal cost**. 'Non-exclusive' and 'non-rival' mean the good is available to everyone and that one person's consumption of the good does not reduce the amount available to others. Finally, 'zero-marginal costs' means the cost of supplying an additional unit of the good to any particular individual is zero.

Examples of pure public goods are national defence, biodiversity, clean air, flood protection, noise and visual amenities. A distinctive characteristic of a pure public good is that consumers do not have the option of not consuming. As suggested earlier, a pure public good will be under-provided because the owner cannot appropriate the full benefits.

Open access and common property goods

Open access goods are rival in consumption, non-exclusive, non-transferable, and often non-enforceable even when ownership rights exist. Examples of open access goods are ocean fisheries and migratory wildlife. Common property goods (e.g., common grazing land) are rival in consumption and are exclusive for a group of people (e.g., a local community). Rights of use may be transferable by individuals or the group. There may not be legal or formal title to ownership but the group may be able to enforce their ownership rights by means of social sanctions.

According to Hardin's 'tragedy of the commons', under open access or common property rights regimes, the resource will be overexploited (Hardin, 1968). However, the point is made in Chapter 11 that under some forms of common property systems (e.g., customary marine tenure), resource management is likely to be more efficient because it is based on communal rules and customs.

Semi (or quasi) public goods

Semi-public (or quasi public) goods are non-rival in consumption, have a zero marginal cost of provision and are non-exclusive although ownership rights exist. An important distinction of semi-public goods is that even though the owner or the providers of the service cannot exclude others from consumption, consumers can choose not to consume. Examples of this category of goods include radio and TV broadcasts and a lighthouse. In the case of radio and TV broadcasts, the signal strength does not depend on the number of consumers (i.e., zero marginal cost). Consumers cannot be excluded (although with cable TV, signals can be encoded), and they may choose not to receive signals by turning of their TV or radio sets. In the case of a lighthouse, ships cannot be excluded from using the beam, although some ships may choose to ignore it.

Congestion goods

Congestion goods are exclusive and and can be either non-rival or rival in consumption. Such goods do not fall neatly into any of these categories and may exhibit characteristics of private goods or public goods at different levels of consumption. An example of this type of good is a campsite (Figure 4.2). At the level *0b*, a certain number of people can enjoy the amenities without reducing other peoples' enjoyment. At this point the marginal cost of an additional person using the amenity is zero and the good exhibits the characteristics of a public good. After point *b*, congestion sets in and the marginal cost becomes positive. After point *c*, the marginal cost (MC) tends to infinity as congestion reaches a maximum. Examples of congestion goods include roads, bridges, an art gallery, fishing and boating sites, and a historic site.

Figure 4.2 An example of a congestion good

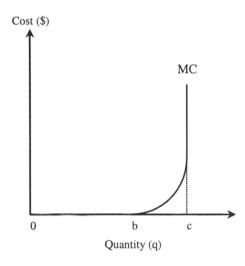

4.2.3   Market Demand for a Public Good

Individuals consume different amounts of a private good. It was explained in the previous chapter that the market demand for a private good is derived by a horizontal summation of the individual demand curves. However, in the case of pure public goods, individuals consume equal amounts. Therefore the market demand curve is obtained by **vertical summation** of the individual demand curves.

Suppose we wanted to estimate the demand for air quality. Let us use abatement of CO_2 emissions as a proxy for air quality. Given that air quality is a non-market good, there would be no market prices and therefore we would have to ask consumers their WTP for given percentage reductions in CO_2 emissions. Table 4.1 presents the data for this hypothetical survey. For each given level of CO_2 abatement, individual A's demand is added to B's demand to obtain the market demand. Let the inverse demand curves be defined as follows:

A's demand:	$p_1 = 10 - 0.05q^D$
+ B's demand:	$p_2 = 20 - 0.1q^D$

$$= \text{Market demand:} \qquad p = 30 - 0.15q^D$$

Table 4.1 A hypothetical demand schedule for air quality
(CO_2 abatement)

Quantity of air quality demanded (% CO_2 reduction) (q^D)	A's WTP for air quality ($) ($p_1 = 10 - 0.05q^D$)	B's WTP for air quality ($) ($p_2 = 20 - 0.1q^D$)	Market demand price for air quality ($) ($p = 30 - 0.15q^D$)
0	10.00	20.00	30.00
5	9.75	19.50	29.25
10	9.50	19.00	28.50
15	9.25	18.50	27.75
20	9.00	18.00	27.00
25	8.75	17.50	26.25
30	8.50	17.00	25.50
35	8.25	16.50	24.75

Notice that for each given level of CO_2 abatement, individual *B* is willing to pay more than individual *A*. This difference could be explained by *B* having a higher level of demand as a result of a higher income or being more environmentally aware. However, there is the likelihood that *B* may understate his or her true preference for the good. This problem is referred to as **strategic behaviour**, or the **free-rider problem**, and is discussed in Chapter 5.

Figure 4.3 presents graphs for the individual and market demand curves for air quality. For a given point on the market demand curve, the corresponding price is given by the vertical summation of the individuals' willingness-to-pay for that level of demand.

4.3   Externalities

In this section, we formally define an externality and outline the causes of externalities. We then proceed to discuss different types of externalities.

Figure 4.3 A hypothetical demand curve for air quality (% CO$_2$ abatement)

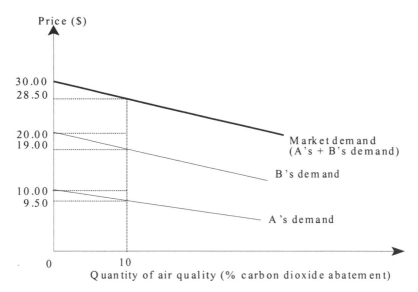

4.3.1 Definition of an Externality

As the word suggests, an externality is an effect that is 'external' to the causing agent. That is, the person causes an effect that impacts on other people. An externality is said to exist when the utility of an economic agent is affected by the actions of another.

An externality is often **negative** (also referred to as an **external diseconomy** or **external cost**). This occurs when the affected person suffers a loss in utility that is uncompensated. Examples of negative externalities are air, water and noise pollution. A **positive externality** (**external economy**, or **external benefit**) occurs when the effect is beneficial to the affected person. An example of a positive externality is immunisation. Immunisation of people in a population helps to reduce the outbreak of an epidemic and therefore benefits those who are not immunised. Another example of a positive externality is where one firm's technological breakthrough benefits other firms in the industry who have not contributed to the research costs.

4.3.2 Causes of Externalities

The following factors give rise to externalities:

1. Interdependence between economic agents: the activity of one agent affects the utility or production function of another. However, the market system fails to 'price' this interdependence, as a result of which the affected party is uncompensated;
2. Lack of or weak property rights: due to lack of or weak property rights, the affected party is unable to demand that the externality be reduced or ask for compensation.
3. High transactions costs: the cost of negotiating, implementing and enforcing an agreement between the parties may be high.

If the affected agent is compensated for his or her loss of welfare, the externality is said to be **'internalised'**, and society is better off by the gainer compensating the loser.

4.3.3   Classification of Externalities

Externalities can be classified in a variety of ways. For example, externalities can be grouped into

- Relevant externalities
- Pareto-relevant externalities
- Static or dynamic externalities, and
- Pecuniary externalities

Relevant externalities

An externality is not relevant so long as the affected person is indifferent to it. It becomes relevant when the affected person is made worse off by the activity and wants the offending person to reduce the level of the activity. For example, if the chicken farm in my backyard interferes with the satisfaction you derive from sitting in your front porch, then it is a relevant externality. On the other hand, if the high decibel music from my stereo does not bother you then the externality is not relevant.

Pareto-relevant externalities

A Pareto-relevant externality exists whenever its removal results in a **Pareto improvement**. A Pareto improvement is a situation where it is possible to take action such that the affected person is made better off without making the offending person worse off. This means that when the level of an externality is optimal, it becomes Pareto irrelevant. To give an example, suppose a telephone company erects a tall transmission tower near a forest. The tower is unsightly and reduces the scenic value of the forest. In this case, a Pareto-relevant externality exists because it is possible for the telephone company to paint the tower in colours that would blend with the foliage. Such a strategy would not interfere with the functioning of the tower and at the same time minimise the impact on the scenic values.

Static and dynamic externalities

To illustrate static and dynamic externalities, take the example of two fishers who are operating under an open access or property rights regime. A static externality is when one creates an externality for the other by overfishing. However, the externality can become dynamic if the offending party is harvesting fish that may have some future value. This could happen, for example, if the offender is harvesting juvenile fish species. In this case, the opportunity cost of the fish reflects a forgone value in the sense that there will be adverse impacts for the future.

Pecuniary externalities

This is a form of externality that is transmitted through the price system. An externality is usually an 'unpriced' effect. However, a pecuniary externality occurs when the externality is transmitted through higher prices or reduced costs. An example of the latter is when a large business moves into a residential area and immediately drives up rental prices. The rent increase creates a negative effect for all those who pay rents and therefore causes a negative externality. An example of the latter is when a manufacturer benefits as a result of a supplier reducing the cost of a product. In a strict sense, the externality in both cases is not a result of market failure. For example, in the first case the resulting increase in rents reflects the scarcity of rental units. Pollution, on the other hand, is not a pecuniary externality because the effect is not transmitted through higher prices. In many cases, the penalties that polluters pay do not reflect the amount of damage inflicted on the environment.

4.4   Type of Market Structure

The type of market structure or organisation can also cause market failure. We consider the following two cases: a perfectly competitive market with external costs and a monopoly.

4.4.1   Resource Allocation in a Perfectly Competitive Market

Consider a gold mining company that dumps mine tailings into a nearby river without paying for clean up or treating the waste. In this case, production at the mine includes the production of gold as well as pollution. Or to put it differently, the river water is an unpriced input in the gold production process.

Let us define the following variables: D = demand curve for gold; MC_p = **marginal private cost** of producing gold (i.e., the firm's supply curve); MSC = **marginal social cost** (Figure 4.4). We assume that MSC is greater than MC_p at any level of output because society considers both the costs of pollution as well as gold production, but the company considers only its marginal private cost. The marginal social cost of gold production is therefore given by the **marginal external cost**, the cost of disutility caused by the externality, plus the marginal private cost. That is:

$$C = MC_p + MEC \qquad\qquad (4.1)$$

Under a perfectly competitive market structure, the company maximises its producer surplus by producing q_0 units of gold. However, from society's point of view, q_0 is not the efficient allocation. Society's net benefits could be maximised by producing less gold, that is, q^* units. The triangular area *abc* represents a deadweight loss to society.

Note the following observations about Figure 4.4:

1. The socially optimal level of pollution (which is assumed to be proportional to gold production) is not zero. This implies that it may not be socially optimal to have zero pollution.

2. In a perfectly competitive market where pollution is unpriced (i.e., there is no pollution abatement), production results in more output than is socially desirable, resulting in excessive pollution.
3. If pollution abatement is enforced, the company will reduce pollution but raise the price per unit of output, resulting in reduced output of gold. However, in this case, the reduced output is the socially efficient level and the higher price is also the efficient price.

Figure 4.4 Resource allocation in a competitive market with externalities

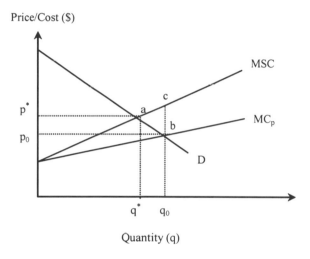

Quantity (q)

4.4.2 *Resource Allocation in a Monopoly*

Assuming a perfectly competitive market and a system of private property rights, the price mechanism will combine to result in an efficient allocation of resources. However, the presence of monopoly rights causes market failure or inefficient allocation of resources from society's point of view. Take the case of a single monopolistic firm with a marginal cost curve, MC, facing a market demand curve, D (Figure 4.5). Under perfect competition, q^* units of the good will be supplied by setting: price = marginal revenue (MR) = marginal cost (MC).

Note, however, that in the case of a monopoly, the demand curve is above the marginal revenue curve and therefore price is not equal to marginal revenue.

Figure 4.5 Resource allocation in a monopoly

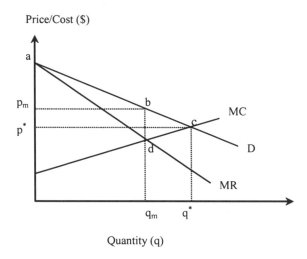

Monopoly profit is maximised by setting *MR* equal to *MC*. This results in less output, q_m, and a higher price, p_m. Consumer surplus under a monopoly is ap_mb, which is less than consumer surplus under perfect competition, ap^*c. Recall that the demand curve *(D)* represents the marginal benefit of goods to consumers. Figure 4.5 indicates that for a monopolist, marginal benefit exceeds marginal cost and therefore the level of output (q_m) is inefficient. Consequently, there is a deadweight loss to society represented by triangle *bdc*.

The monopolist's production decision may be somewhat unexpected because it seems to suggest that less pollution will be created in a monopoly than in perfect competition. Furthermore, the monopolist's higher initial price (p_m) suggests that, given a fixed stock of natural capital, the price path will be less steep over time than in perfect competition (Pearce and Turner, 1990). However, caution must be exercised in making such comparisons because other factors (e.g., elasticity of demand) affect the outcome.[19]

---

[19] Stiglitz (1976) argues that the extraction and price paths may be identical for monopoly and perfect competition, given a constant elasticity of demand for the resource.

4.5   Policy (Government) Failure

Policy (or government)   failure occurs when the government creates incentives for the prices of certain goods and services to be lower than the actual cost of production per unit. An example of policy failure is a government subsidy on pesticides which provides incentives for farmers to use more pesticide than is socially efficient, resulting in adverse environmental impacts. This type of subsidy is referred to as an input subsidy and other examples include energy, fertilizer and irrigation subsidies. Other types of subsidies include guaranteed prices for agricultural products and subsidies for land clearing which tend to encourage large scale production and loss of forest cover. Sometimes, the subsidies could be put on consumer products. In this case, if the particular product, or the input used in its production, is environmentally damaging, then the net effects on the environment could be negative.

Table 4.2 presents rough estimates of subsidies worldwide. Among other things, the figures indicate that subsidies worldwide are substantial, amounting to about 4 percent of the world's GNP. This is estimated to be about US$25 trillion. The subsidies are about 14 times Overseas Development Assistance (ODA) flows in any given year. It is interesting to note that the highest subsidies are in the Organisation for Economic Cooperation and Development (OECD) countries. In general, subsidies in the developing countries are on the decline as most of them have adopted structural adjustment programs over the past two decades.

4.6   Solutions to Environmental Pollution Problems

Two main approaches have been proposed for dealing with externality problems (Figure 4.6). The first approach, known as the property rights or market solution, was proposed by Ronald Coase and involves allowing the free market system to solve the problem through bargaining between the affected parties. The second approach is by means of government intervention. Government pollution control policies can take two main forms: market based instruments (MBI) and command-and-control (CAC) instruments. We begin the discussion with the Coasian or market solution.

Table 4.2 Provisional estimates of world subsidies (US$ billion)

Sector	OECD	Non-OECD
Water:		
Irrigation	3+	20
Sanitation	?	5
Supply	?	28
Fossil Fuel	34–52	100+
Nuclear Power	9–14	0?
Agriculture:		
Pesticides	0?	2?
Fertilizers	0?	4
Outputs	336	0?
Transport	79–108	?
Total	**461–513**	**159**

Source: Pearce and Ozdemiroglu (1997).

4.6.1   *The Property Rights or Coasian Solution*

According to the **Coase Theorem** (Coase, 1960)[20], negotiation between two parties involved in an externality will eliminate Pareto-relevant externalities and result in an efficient solution if property rights are well specified. The final allocation does not depend on the initial assignment of property rights and the only effect is the distribution of costs and benefits. This theorem is illustrated with the aid of Figure 4.7.

Consider two parties, one is a factory which is polluting a nearby river with industrial effluent. The other party is a community which utilises the river water for drinking purposes. The community has a downward-sloping demand curve for pollution abatement. This is referred to as the marginal benefit (MB) curve because it indicates the consumer's benefits from consuming an additional unit of clean water.

---

[20] Ronald Coase won the 1991 Nobel prize in economics for his use of property rights regimes to highlight institutional issues and the functioning of the economy.

Figure 4.6 Pollution control instruments

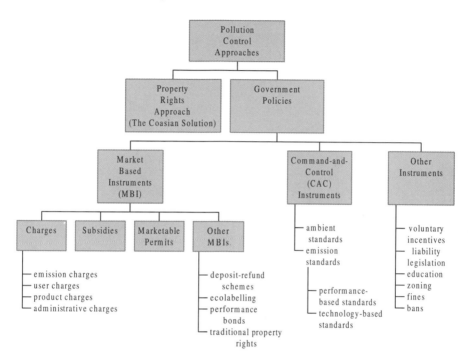

The factory has an upward sloping supply curve for pollution abatement which reflects the marginal cost of increasing levels of pollution abatement. This curve is also the marginal external cost (MEC) curve. Note that in the absence of a legal requirement to abate pollution, the factory has an incentive to supply zero pollution abatement (q=0%) because this is the point at which profits will be maximised.

According to the Coase Theorem, the socially optimal level of pollution abatement will be $q^*=60\%$. Let us first consider how this solution is achieved in the case where the community has the property rights to the river.

Figure 4.7 The Coasian solution to the externality problem

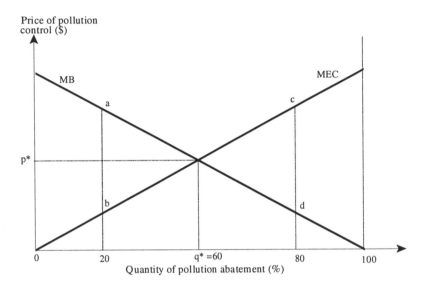

Case 1: The community has the property rights

Assuming the community has the property rights to the river, it would like to have zero pollution or 100 percent pollution abatement. Therefore the starting point will be *q=100%*. Let us bear in mind that the downward-sloping demand curve for pollution abatement implies that at zero percent pollution abatement the community's WTP for pollution abatement is initially high. However, as the units of pollution abatement increase the community's WTP declines. Thus, at the level of *q=100%* the community's WTP for pollution abatement is lower than the polluter's marginal cost of supplying pollution abatement. There is therefore a possibility for a trade.

To illustrate, let us consider the point *q=80%*. At this level, the maximum amount the polluter is willing to pay to supply an additional unit of pollution is *c*, which is higher than the minimum compensation, *d*, that the community will demand per unit of pollution abatement. In this particular case, the factory would be willing to offer compensation of up to *cd* per unit to the community to induce them to accept less pollution abatement. The community would be willing to accept this amount because even though it suffers a welfare loss from having less pollution abatement, this is offset by the compensation which exceeds their minimum demand price of *d*.

The move from *q=100%* to *q=80%* is a Pareto improvement because at least one party is better off and no one is worse off. The factory could negotiate less and less pollution abatement. However, it would not offer a level of pollution abatement less than $q^*=60\%$. This is because below $q^*=60\%$ the minimum compensation demanded by the community (*a*) exceeds the marginal cost of supplying pollution abatement (*b*). Therefore, the factory will choose to supply pollution abatement.

Case 2: The factory has the property rights

When the factory has the property rights to the river, the starting point is *q=0%* because it has the right to pollute. However, once again there is potential for a trade because the community's WTP for pollution abatement exceeds the factory's marginal cost of pollution abatement. Suppose the community wishes to increase pollution abatement to *q=20%*. It could offer a 'bribe' of *ab* per unit to the factory to induce it to supply more pollution abatement. The factory would be willing to accept this amount because it exceeds the marginal cost of supplying pollution abatement at that level. The factory has no incentive to provide pollution abatement beyond $q^*=60$ because the marginal cost of supplying pollution abatement exceeds the maximum unit bribe the community is willing to offer.

From the foregoing, it can be seen that, irrespective of who has the property rights, equilibrium is achieved at a quantity of q^* and a price of p^*. The outcome of this market solution is an efficient allocation of resources and the removal of the Pareto-relevant externality. However, the distribution of costs and benefits is different in each case. For example, when the offending party has the property rights, it is the affected party who makes the payment, and vice versa.

Limitations of the Market Solution

The Coasian analysis has made a number of contributions to our understanding of externalities and the concept of property rights, in particular, trade in rights. However, the theorem is based on the following key assumptions that may not apply in the real world.

- zero transactions costs

- well-defined property rights
- perfect competition, and
- no income (or wealth) effects
- no free-rider effects

Zero transactions costs

Transactions costs refer to expenses that are incurred in the process of negotiating. Examples include legal fees, the cost of organising and bringing the parties to the bargaining table, and so on. These costs can be high especially when the population of affected people is large and scattered. These costs can be large enough to discourage some parties from seeking a negotiated settlement to the externality.

Well-defined property rights

The Coase Theorem implicitly assumes that either one of the two parties has a complete set of property rights. However, in many cases, property rights are either poorly defined or non-existent. Lack of property rights is probably one of the most significant obstacles to bargaining. In large resource development projects, it is often difficult to know precisely who the landowners are. This opens the way for speculators and activists to influence the negotiations.

Perfect competition

The Coasian solution assumes a perfectly competitive market. The affected party's 'bargaining curve' is its marginal benefit curve, from which it can determine how much to pay and how much compensation to demand. Likewise, the offending party's 'bargaining curve' is its marginal external cost curve. Under these conditions, the equilibrium solution is obtained by setting price equal to marginal cost. In the absence of perfect competition, the factory will set marginal revenue (and not price) equal to marginal cost, in which case q^* is no longer the optimum solution.

Income effects

The theorem assumes that there are no wealth or income effects. However, the assignment of property rights to one party, in reality, results in an income transfer to that party. The equilibrium level of pollution abatement will be more than q^* when the affected parties have the property rights, and less than q^* when the offending party has the property rights. It is also possible that when the offending party has the property rights, it may be tempted to increase pollution so as to increase the size of the bribe.

Free-rider effects

When there are many affected parties the free-rider problem may make it difficult to negotiate an efficient solution. An example of free-rider effects is when several communities have to negotiate with a big multinational corporation over resource development. It has been suggested that in such cases the government could facilitate pollution abatement by means of legislation and regulation. We discuss government policies in the following section.

4.7 Government Policies

The Coasian solution relies on negotiation between the parties to eliminate the inefficiency created by the externality, upon specification of property rights. However, as we have seen above, there are various reasons why the Coasian solution may not work in practice. There is therefore a need for government policies to correct the externality problem. As shown in Figure 4.6, there are two main categories of government pollution control instruments: market-based instruments and command-and-control instruments. Market-based instruments include charges, subsidies, marketable permits, and deposit-refund schemes. The CAC approach, also known as standards or regulation, includes ambient standards, performance-based standards, and technology-based standards. We begin the discussion with CAC approaches, which are, by far, the most commonly used approach.

4.7.1 Command-and-Control Systems

Command and control mechanisms are the oldest forms of pollution control policies in existence. As the name implies, the CAC mechanism consists of a 'command', which sets a standard (e.g. the maximum level of pollution allowable), and a 'control', which monitors and enforces the standard. In general, there are three types of standards: ambient standards, emissions standards and technology standards (Figure 4.8).

Ambient standards

Ambient standards set the minimum desired level of air or water quality, or the maximum level of a pollutant, that must be maintained in the ambient environment. For example, an ambient standard for dissolved oxygen in a certain river may be set at 3 parts per million, meaning that this is the lowest level of dissolved oxygen allowed. Ambient standards are normally expressed in terms of average concentration levels over some period of time. In general, they cannot be enforced directly. What can be enforced are the emissions from various sources that can lead to ambient quality levels.

Emission standards

Emission (or effluent) standards specify the maximum level of permitted emissions. They can expressed in terms of quantity of material per unit of time (e.g., grams per minute, or tons per week), total quantity of residuals, residuals per unit of output (e.g. SO_2 emissions per kilowatt-hour), or residuals per unit of input (e.g., sulphur content of coal used in power generation). Emissions standards are also referred to as performance standards because they refer to end results to be achieved by the polluter.

Technology standards

Technology-based standards specify the technology, techniques or practices that a firm must adopt. This type of standard could be in the form of 'design standards' or 'engineering standards'. Other related types of standards include product standards that specify characteristics that goods must have, and input standards that specify conditions that inputs used in the production process must meet. Technology-based standards not only specify emissions limits, but also the 'best' technology that must be used. The main difference between a performance standard and a technology standard is that the former sets a constraint on some performance criteria and then allows individual firms to choose the best means of achieving it. On the other hand, the latter actually dictates certain decisions and techniques that should be used to achieve the criteria.

Referring to Figure 4.8, the socially optimum standard and penalty will be q^* and p^*, respectively. Let the actual standard applied be represented by the vertical line, q^s. The usual practice is to impose a penalty for polluting beyond the standard. Given the penalty, p, the efficient level of pollution abatement for the firm will be q, and there will be a deadweight loss to society of *abc*.

Although standards are widely understood they have serious deficiencies in terms of providing incentives to reduce pollution. Standards have a number of disadvantages. First, firms have no incentives to reduce pollution beyond the standard. Second, penalties for violating standards tend to be too low and enforcement tends to be weak. Third, to set an optimum standard and penalty, the government must know the demand (marginal social benefit) and the supply (marginal social cost) curves for pollution abatement. However, since air (or water) quality is a non-market good, the demand curve is not directly observable[21]. Also, it is difficult for the government to know exactly the industry's marginal abatement (or external) cost curve, given the large number of polluters. In light of the above reasons, it is likely that the government will set the standard at a point other than q^*.

Figure 4.8 An emission standard

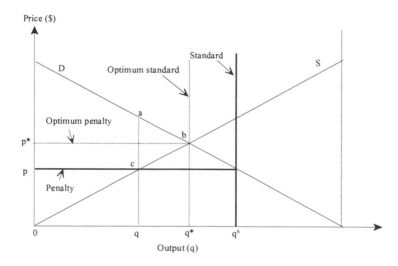

Fourth, to be effective, standards need to be revised frequently in response to rapidly changing circumstances. However, in practice, legislation tends not to keep up with the pace of change. Fifth, the financial costs of setting

---

[21] Chapter 5 reviews methods for estimating the value of non-market goods.

standards will be high if there are many polluters. These costs include the administrative cost of implementing the system of standards and the monitoring and enforcement costs. Six, there could also be political costs if the standards are stringent and businesses are adversely affected.

By far the most serious defect of standards is the fact that they are uniformly applied to all firms and regions. This takes flexibility away from polluters. The fact of the matter is that pollution abatement costs differ between firms and regions, and forcing high-cost abaters to reduce pollution as much as low-cost abaters, results in more resources being used to achieve a cleaner environment. The community can achieve cost savings by having more abatement undertaken by firms who can do so at a relatively lower cost.

On the positive side, standards are a more widely understood form of environmental policy. They are more pragmatic and, perhaps, more socially acceptable than MBIs, especially when there is the possibility that a pollutant will affect human health. A good example is the disposal of nuclear or human waste. In this regard, standards are considered to be consistent with the **Precautionary Principle** (Ramchandani and Pearce, 1992) which takes account of intergenerational and intragenerational concerns.[22] Finally, the political costs of standards are lower compared to market based instruments such as taxes and subsidies as setting standards does not incur direct budgetary implications.

4.7.2   Market-Based Instruments (MBIs)

Market-based instruments use price or some other economic variables to provide incentives for economic agents to abate pollution. As described in Figure 4.6, MBIs include **charges**, **subsidies**, **marketable permits**, **deposit-refund schemes**, and **performance rating schemes**.

Charges

A charge or tax is, in effect, a negative price that is levied in proportion to the amount of pollution. This type of tax is also referred to as a **Pigovian tax** after the British economist, Arthur Pigou, who proposed the idea in 1920 (Pigou, 1920).

---

[22] The issues of intergenerational and intragenerational equity are discussed in Chapter 11 in relation to sustainable development.

There are at least four types of charges: **emission charges, user charges, product charges** and **administrative charges**. A product charge is a variation of emission charges whereby taxes are levied on goods produced with polluting inputs. An example of a product charge is a carbon (fuel) tax. User charges are fees levied for using an environmental resource. They are often not directly related to the level of pollution but rather aim to recover some portion of the abatement cost. Finally, administrative charges are service fees levied by a government authority for implementing or monitoring a regulation. Charges are based on the **'Polluter-Pays Principle (PPP)'** which asserts that the polluter should bear the cost of any abatement taken to maintain an acceptable level of environmental quality (OECD, 1989).

To give an example of a Pigovian tax, suppose the demand curve for pollution abatement is D and the supply of pollution abatement is S (Figure 4.9). The government would try to set the charge at p^*, where demand equals supply. However, this approach faces a similar constraint as standards in the sense that both the demand and supply curves are not known with certainty. In practice the charge is likely to be set at a point $p < p^*$ at which the level of pollution abatement will be less than optimum.

Pearce and Turner argue that attempting to calculate an 'optimal' tax is unrealistic. What is needed is 'the kind of information that would tell us whether we are very wide of the mark in taking a particular pollutant or whether we are in the right ballpark (Pearce and Turner, 1990:97).

Economists prefer charges to other pollution abatement alternatives because charges offer firms an economic incentive to reduce pollution. Different firms have different pollution abatement costs. By imposing a charge per unit of pollution, emission charges induce firms to lower their emissions to the point where the marginal cost of abatement equals the charge. Unlike standards, which are applied uniformly to all polluters, charges enable firms to adopt a cost-effective solution to pollution abatement. Compared to standards, there is a stronger incentive for firms to adopt new technology in order to lower the charges they have to pay.

In spite of the above advantages, charges and taxes do have some disadvantages. The problem associated with setting an optimum tax due to uncertainty about the demand and supply curves has already been discussed above. In addition, the costs of monitoring emissions could be high. Furthermore, there are a number of equity issues arising from such a tax. First, firms could pass on a portion of the tax to consumers in the form of higher product prices.[23] If the product is a necessity, then low-income households who spend a high proportion of their income on such goods would be adversely affected.

Figure 4.9 An example of a Pigovian tax

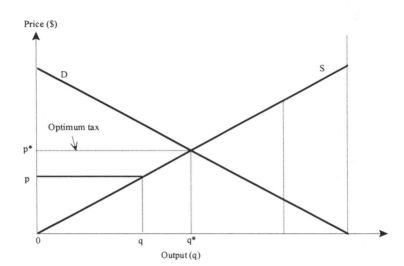

Second, imposing a tax could also lead to job losses as firms minimise their costs in order to increase pollution abatement.[24] Finally, the costs of monitoring compliance may be high if the charges are based on the emissions and the number of pollutants is high.

Subsidies

An alternative to taxes is for the government to subsidise the polluter per unit of reduction in the level of pollution. The subsidy could also be offered for the purchase of pollution abatement equipment or technology. Referring to Figure 4.9, the amount of the subsidy will be p^* per unit of pollution

---

[23] See Exercise 5, Chapter 3 for a problem based on this issue.
[24] See Section 3.7.2 for an analysis of the employment effects of an emission standard.

abatement. In this case, the optimum level of pollution abatement will be q^* units.

In theory, both taxes and subsidies should result in the same optimum level of pollution abatement. However, if there is unrestricted entry into the industry subsidies could attract more producers. Therefore, in the long run aggregate pollution could increase under subsidies but decrease under charges. Also, subsidising pollution abatement may be seen as socially 'unjust' because what it effectively does is to redistribute income away from society to polluters. In general, subsidies are politically attracted to the few people who tend to receive them.

Marketable Permits

Marketable (or tradeable) pollution permits are a relatively new approach to controlling environmental pollution. Marketable permits were first introduced in the U.S. in 1977 as part of the Clean Air Act. Under this system, the government issues a fixed number of permits or 'rights to pollute' equal to the permissible total emissions and distributes them among polluting firms in a given area. A market for permits is established and permits are traded among firms. Firms that maintain their emission levels below their allotted level can sell or lease their surplus allotments to other firms or use them to offset emissions in other parts of their own facilities. A variant of marketable permits is the **Individual Transferable Quota** (ITQ) system that has been used in Australian and New Zealand to manage fisheries (Box 4.1). It was first used in New Zealand in 1986, and has since been used in Australia to manage the Southern Bluefin tuna fishery and in the U.S. to manage the Atlantic surf clam industry. A new form of MBI that is similar to marketable permits is **Offset-Banking** (Box 4.1).

To demonstrate how permit prices are determined, let the socially acceptable level of emissions in a given geographical area be q_a (Figure 4.10). The supply $(\overline{S})$ of permits is therefore fixed at q_a. The demand for permits, in this, case is equal to firms' marginal abatement costs (MAC). The price of a permit, p^*, is determined at the point of intersection between the demand (i.e., MAC) and the vertical supply curve. After the aggregate level of emissions and the permit price have been determined, firms are then free to trade in permits. Firms with MACs below the permit price, p^*, have an incentive to sell permits to those with MACs above the permit price.

Box 4.1 Individual transferable quotas in fisheries management:
the case of New Zealand

In an effort to combat the problem of declining fish stocks, the New Zealand government in 1983 introduced an ITQ system. The ITQ system is a form of marketable permit system. In the case, the government freely distributed quotas among the industry participants based on historical catch. Each quota allowed the owner to harvest a certain quantity of fish to be harvested. The total tonnage associated with the quotas was made to equal the total allowable catch (TAC) that had been determined by the government to be sustainable. The Fisheries Board was empowered to monitor the fish catch through sales to retailers. Initial assessment indicated that there was an improvement in fish stocks. However, in 1989, it was found that the initial distribution of ITQs was excessive and the government decided to reduce the TAC. Despite some teething problems, the ITQ system has proven to be an effective tool to manage fisheries resources in New Zealand's exclusive economic zone.

Source: Based on Clark (1991).

The flexibility in allowing trade in permits makes the marketable permit system a more cost effective way of getting firms to comply with a given standard compared to the the CAC approach (Tietenberg, 1985).

Marketable permits have some advantages and disadvantages. The advantages include:

1. The permit system is transferable. Innovative firms can profit from selling their permits. There is therefore an incentive for firms to invest in pollution abatement.
2. The system makes allowance for industrial development. New polluters are able to purchase unwanted permits from established firms. Both new and old firms are encouraged to acquire efficient pollution abatement capabilities.
3. The system can generate revenue for the government, although the cost of enforcing penalties must also be considered. The income generated from the scheme could be considered as a form of compensation to the public.

Box 4.2 Offset banking

Offset banking is a new form of MBI that is in a way similar to permit trading which has good prospects for use in urban areas. The basic concept underlying offset banking is that if a developer is proposing a project that will create additional pollution in a given location, it must first 'offset' this increase by reducing pollution elsewhere. For example, the developers of a new golf course might be required to fund best management practices in nearby agricultural areas in addition to on-site best management practices. Or a new housing development might only be approved if, as part of the development, septic tanks in an existing development are installed. The basic intent of offset banking is to ensure that development proceeds without any net environmental impacts.

An offset bank can be privately or publicly owned. It is not a bank in the usual sense of the word. Rather, it involves completion of one or more projects in which environmental remediation is carried out. By completing these environmental projects, the offset bank earns 'credits' that can then be sold to developers who are creating net-impacts on the environment.

Offset banking has been applied in the United States. Examples include wetland mitigation banking (US Corps of Engineers *et al.*, 1995), mitigation of streambank impacts, and in controlling air quality impacts. The main advantage of offsets is that they can generate substantial cost savings for businesses compared to having to fully mitigate any impacts on-site. Thus, the system has the potential to reduce environmental risk and at the same time encourage investment.

Source: Morrison (2003).

The disadvantages include the following:

1. The market for permits may not be perfectly competitive if the number of polluters is small. In this case, the bigger firms may be able to exert some market pressure on permit prices.
2. The scheme may involve high transactions costs such as administrative, monitoring and enforcement costs. Huber *et al.* (1998) suggest these costs could be higher than under the CAC (see Box 4.3). This is due to the fact that the monitoring may be as much as in CAC, but there would also be additional costs.

3. Marketable permits are difficult to operate when there are several pollutants in the area. With several pollutants, it is more difficult to measure aggregate emissions and to monitor compliance.
4. There have been low levels of trading permits in some of the U.S. systems.

Figure 4.10 Marketable permits

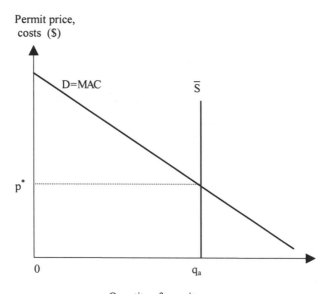

Quantity of permits, q

Other MBIs

Other types of market-based incentive instruments include deposit-refund schemes, ecolabelling and performance rating, performance bonds, and traditional property rights.

Deposit-Refund Schemes

As the name suggests, a deposit-refund scheme involves a 'deposit', which is a front-end payment for a potential polluting activity, and a 'refund', which is a guarantee of a return of the payment upon proving that the polluting activity did not take place. Deposit-refund schemes have some of the economic incentive characteristics of charges. The 'deposit' is an attempt to

---

Box 4.3 Exploding a myth: MBI or CAC?

A common assumption regarding MBIs is that they form a ready substitute for outdated or inefficient CAC regulatory procedures. At least, in the case of Latin America and the Caribbean, the implementation of MBIs will not provide a quick panacea for the problems associated with CAC approaches. Although there is definite scope for the gradual implementation of MBIs, there are significant constraints in three areas.

1. Institutional constraints: in most countries, institutional weaknesses (e.g., underfunding, inexperience, or lack of political will) limit the effective implementation of MBIs.
2. High administrative costs: monitoring requirements and other enforcement activities associated with CACs not only apply to MBIs, but also there are additional costs related to the required institutional changes.
3. Macroeconomic reforms may hinder or help MBI implementation. For example, deregulation often strikes down regulations necessary for MBIs to operate, and 'downsizing' of the public sector may limit the amount of institutional strengthening required for MBIs to be effective.

Source: adapted from Huber *et al.* (1998).

---

capture the marginal external cost of improper waste disposal. The consumer is forced to account for the external cost of pollution by making an up front payment. However, the 'refund' is a reward to properly dispose of waste. The deposit-refund scheme is one type of MBI that is well established in both advanced and developing countries. In the latter, collection of paper, plastic and other recyclable material is a source of informal occupation for unskilled workers.

Deposit-refund schemes have the following advantages:

1. It is a voluntary system that tries to change environmental behaviour at least cost to the government. The monitoring and enforcement costs are minimal because it requires limited supervision.

2. The system can be used to encourage recycling and more efficient use of raw materials. That is, the 'deposit' is a tax that encourages firms to use raw materials more efficiently during the production process. On the other hand, the 'refund' encourages them to properly dispose of the waste products.

Ecolabelling and Performance Rating

In this approach, the government supports an ecolabelling or performance rating program that requires the firm to provide information on the final end-use product in order to give consumers more market information. The performance rating is on the basis of ISO 14000 voluntary guidelines that includes the following: zero discharge of pollutants, adoption of pollution abatement technology, submission of mitigation plans. Ecolabels are attached to products that are judged to be 'environmentally friendly'.

Performance Bonds

The performance bond system requires a company to deposit a certain amount of money in the form of a bond before commencing a project. The money is returned to the company if it complies with regulations concerning rehabilitation or clean up of the project site after the project is completed. Performance bonds vary in the terms and conditions of the penalty clauses. In some cases, the bond forms part of a permit issued by the relevant government agency, setting out the type of activity allowed and the location of the activity.

Traditional Property Rights

It has been shown that integrating traditional or customary communal rights systems with western systems of natural resource management can assist in the process of sustainable development (see case study below). In the past, resource managers in developing countries have concentrated their management efforts on licenses and quotas in order to prevent overexploitation of natural resources such as fisheries and forests. In many cases, these efforts have been unsuccessful. Customary communal rights systems, whereby local people manage their own resources through the establishment of private or communal ownership over common property resources, have better chances of success. This is because the 'owners' of the resource have an interest in its current and future productivity and would be inclined to control exploitation so as to maximise the net benefits.

Traditional property rights have been used to manage coastal fisheries in various Pacific island countries such as Papua New Guinea, Fiji, and Vanuatu (see Appendix 4.1).

4.8   Choosing the Right Policy Instrument

Before an appropriate instrument can be chosen, each country needs to carefully consider the following issues: (i) is the instrument economically efficient? (ii) is it effective (or dependable)? (iii) is it adaptable (i.e. flexible)? (iv) is it equitable?, and (v) is it politically acceptable?

4.8.1 Economic Efficiency
Efficiency or 'allocative efficiency' refers to the use of society's resources in an optimal way, i.e. without any wastage or additional costs. For a policy to be considered 'efficient', the total costs (including costs to the government, individuals and firms) involved in implementing the policy must outweigh the total benefits. Thus, improved efficiency is associated with cost savings to firms or net benefits in terms of improvement in environmental quality and natural resource stocks. For natural resources, examples include increases in sustained yields in fisheries and forestry. Policies such as MBIs do have associated costs and these costs must be weighed against any benefits to determine whether they are a worthwhile venture for a specific country. It could very well be that for some countries, MBIs may be too costly due to additional factors such as the possibility of low compliance and costs of installing monitoring equipment.

4.8.2 Effectiveness
'Effectiveness' refers to the degree to which the policy or measure achieves the environmental objective of protecting the environment or natural resource. For example, with regard to a pollution abatement policy, the issue is whether the policy actually results in a reduction of pollution. Different instruments will have different levels of effectiveness and should be compared. For example, in the case of controlling highly hazardous chemicals, CACs may be more effective than MBIs since the former has more certainty in terms of the quantity of abatement. While MBIs may be

more effective with other types of environmental damage (e.g. biodegradable waste).

4.8.3 Adaptability

A good instrument should continue to be effective in the face of changing circumstances (e.g. changes in prices, environmental conditions and technology). In this respect, a policy instrument such as CAC is not adaptable or flexible since the standards are often not revised frequently. Also it does not give firms the incentive to adapt to changing technology.

4.8.4 Equity

An important consideration in assessing policy instruments is the issue of equity or 'fairness'. This has to do with the distribution of the costs and benefits among different groups in the population. Impacts on low-income groups may be a concern, as well as effects on the profitability and competitiveness of local industry. Before a particular policy is adopted, there is the need for a distributional analysis to identify the various groups (consumers and producers) who will be adversely affected. In some cases, it might be appropriate to use revenues raised from the policy to address inequalities in income distribution resulting from the policy.

The OECD has recommended the following checklist for conducting distributional analysis (OECD, 1994).

1. What is the benchmark for comparison of different policy tools?
   The two options here are to compare the new instrument against the old one, or where there is no existing instrument, to compare the instrument with the 'no regulation' case (i.e. the status quo).
2. Will the economic instrument lead to government revenues and to what use will these revenues be put?
3. What are the initial impacts of the instruments?
4. What are the relevant groups for which impacts will be addressed (either quantitatively or qualitatively)?
5. What empirical steps are needed to determine the final impacts (taking into account the shifting of costs and benefits to other groups)?

4.8.5 Political Acceptability

As indicated earlier environmental policy instruments such as taxes tend to be unpopular. Opposition to particular instruments could also stem from lack of knowledge. For example, establishing a system of tradeable permits could be misconstrued as giving people a 'license' to pollute. Thus, one of the ways to increase political acceptability is through a program of public education and consultation. Other ways to promote political acceptability include gradual implementation of the policies and explanation of how any revenues raised would be used.

4.8.6 Concluding Remarks

Market-based instruments have been shown to achieve the same environmental objectives at a lesser cost than CAC mechanisms (Tietenberg, 1991). They can also generate significant revenues for the government. These revenues could then be used to address any equity issues if that is a political concern.[25] Given their limited resources, developing countries would need to consider the introduction of MBIs in the long run in order to maintain their external competitiveness. MBIs may be costly in some countries due to technical reasons, while in others the economic and institutional framework needed to support them are either weak or non-existent. In general, MBIs are more likely to be successful in the region if they are introduced gradually and, where appropriate, in combination with CAC instruments. A mix of CACs and MBIs is also important for the overall success of environmental policy if the institutional/legal framework is poor. There is also the need to educate industries and the general public about MBIs. In particular, it is important to stress that MBIs provide incentives for reducing pollution rather than simply being a tool to punish polluters.

4.9   Summary

In this chapter it has been demonstrated that the market system fails to fully incorporate the value of environmental goods and services because such goods are often not traded in markets and therefore tend to be underpriced or

---

[25] It has been argued that in some cases, the total distributional impact of MBIs can be less than CACs. Evidence from the US suggests that CACs can be regressive. That is, the poor bear a greater burden of the cost relative to their income compared to the rich.

unpriced. Market failure can occur due to the following factors: lack of or weak property rights; public goods and/or common property characteristics; externalities; and type of market structure. When there are weak or no property (or ownership) rights for an environmental good there is a tendency to overexploit it. There are few incentives to conserve such goods because an individual (or group) cannot expect to appropriate all the benefits. Market failure can also occur because a good displays public goods characteristics. A pure public good is a good that is non-exclusive and non-rival in consumption, and has a zero marginal cost of provision. The owner of a public good will not supply the optimal amount because he or she cannot exclude others from consuming it.

The third type or cause of market failure is due to externalities. An externality is said to occur when one person's actions affect the welfare of another, but the affected person is not compensated. Market failure also occurs when, due to lack of (or weak) property rights, the affected person cannot demand compensation from the offender. Finally, market failure can occur when one person exerts monopoly influence in the market place. In this case, the equilibrium price is higher and the equilibrium quantity is less than in a perfectly competitive market. In the case of some non-renewable natural resources, such a situation may actually result in more conservation.

We considered two main approaches to the solution of environmental pollution problems: the market (Coasian) solution and government intervention. The Coasian solution relies on the market system to resolve the externality problem through bargaining between the affected and offending parties. The outcome of the negotiation will be the elimination of the Pareto-relevant externality irrespective of who has the property rights. However, it was explained that the Coasian solution is based on some strong assumptions including zero transactions costs, well-defined property rights, perfect competition and zero income effects. Since all or some of these assumptions may not hold, it is argued that there is a need for government intervention.

We discussed two main types of government pollution abatement instruments: command-and-control (CAC) and market-based incentive mechanisms. The advantages and limitations of these two approaches were addressed. MBIs are generally preferred over CAC approaches because they offer an economic incentive to reduce pollution. However, CAC approaches are more appropriate when there is uncertainty about health effects. Although MBIs are more cost-effective and provide revenue for the government, their administratuve costs could be higher than CAC (see Box 4.1).

In conclusion, the point must be made that each pollution abatement strategy has particular strengths and weaknesses, as well as associated costs. When considering options for dealing with particular pollution problems, the expected benefits must be compared with the costs. In particular, it is important to ensure that a chosen instrument(s) is economically efficient, effective, and adaptable. In order to garner community support, it is also essential to ensure that the instrument is equitable and politically acceptable.

Key Terms and Concepts

ambient standards
characteristics of property rights
charges
command and control
common access goods
congestion goods
deposit-refund schemes
dynamic externality
ecolabelling
emission charges
emission standards
externality
free-rider effects
income effects
marginal external cost
marginal private cost
marginal social cost
market (Coasian) solution
market failure
marketable permits

market-based instruments
open access goods
opportunity cost
pareto optimality
Pareto-relevant externality
pecuniary externality
performance-based standards
polluter pays principle
precautionary principle
product charges
public goods
pure public goods
relevant externality
rival in consumption
static externality
semi-public goods
subsidies
technology-based standards
user charges
vertical summation

Review Questions

1. Explain the meaning of the following economic terms:

- Opportunity cost
- Pareto optimality
- Perfect competition

2. Describe the properties of a fully specified set of property rights.

3. Describe the properties of the following goods:
   - Private good
   - Congestion good
   - Common access good
   - Semi-public or quasi good
   - Pure public good

4. Define the following terms:
   - Externality
   - Pareto-relevant externality
   - Pecuniary externality
   - Precautionary Principle

5. Explain the meaning of the term 'internalising an externality'.

Exercises

1. What is meant by 'market failure'? Explain the causes of market failure.

2. Explain how the Coase Theorem can provide an optimal solution to an externality problem such as pollution. State the assumptions of the theorem and explain why it is unlikely to lead to an efficient outcome.

3. Compare and contrast market-based incentive approaches (e.g., charges) with command-and-control approaches, using the problem of air pollution as a point of reference.

References

Clark, I. (1991). New Zealand's Individual Transferable Quota System for Fisheries Management—History, Description, Current Status and Issues, in T. Yamamoto, and K. Shorts, (eds.), *International Perspectives on Fisheries Management*, Tokyo.

Coase, R. (1960). The Problem of Social Cost. *Journal of Law and Economics*, 3: 1-44.

Hardin, G. (1968). The Tragedy of the Commons. *Science,* 168: 1243-1248.

Huber, R., Ruitenbeek, J. and Serôa da Motta, R. (1998). *Market-Based Incentives for Environmental Policymaking in Latin America and the Caribbean: Lessons from Eleven Countries.* World Bank, Washington, D.C.

Morrison, M. (2003) Offset Banking — A Way Ahead for Controlling Nonpoint Source Pollution in Urban Areas. Paper presented at the inaugural national workshop of the Economics and Environmental Network, Australia National University, 2-3 May, Canberra, ACT.

Organisation for Economic Co-operation and Development, OECD (1989). *Economic Instruments for Environmental Protection.* OECD, Paris.

Organisation for Economic Co-operation and Development (1994). *The Distributive Effects of Economic Instruments for Environmental Policy*, OECD, Paris.

Pearce, D. and Ozdemiroglu, E. (1997). *Integrating the Economy and the Environment: Policy and Practice*, Economic Paper 28, Commonwealth Secretariat, London.

Pearce, D.W. and Turner, R.K. (1990). *Economics of Natural Resources and the Environment.* Harvester Wheatsheaf, Hertfordshire, U.K.

Pigou, A.C. (1920). *The Economics of Welfare.* Macmillan, London.

Ramchandani, R. and Pearce, D.W. (1992). Alternative Approaches to Setting Effluent Quality Standards: Precautionary, Critical Load and Cost Benefit Approaches. WM 92-04, Centre for Social and Economic Research on the Global Environment (CSERGE) Working Paper, University of East Anglia and University College, London.

Stiglitz, J.E. (1976) Monopoly and the Rate of Extraction of Exhaustible Resources. *American Economic Review,* 66: 656-661.

Tietenberg, T. (1985). *Emissions Trading.* Johns Hopkins University Press, Baltimore, Maryland.

Tietenberg, T. (1991). Economic Instruments for Environmental Regulation. In D. Helm (ed.), *Economic Policy Towards the Environment,* Blackwell, Oxford.

Appendix 4.1 Customary Marine Tenure Systems in Papua New Guinea[26]

The traditional model used for fisheries management in Papua New Guinea (PNG), as is the case in most countries, is based on the 'tragedy of the commons' advanced by Hardin (1968). According to this model, if people weigh their private benefits against private costs, they will overexploit common resources when given free access. In terms of fisheries resources, this model predicts that where access is free, there will be intense competition resulting in overexploitation and extinction of the fish stocks. In line with this model, the focus of fisheries management has been to limit fishing effort by imposing rules. Here, it is argued that a system of customary marine tenure (CMT) is a more effective management strategy because it is based on community participation in decision-making and is derived from kinship and lineage structures. This case study is organised as follows. Firstly, the concepts of sea tenure and customary marine tenure are defined. Secondly, we describe the nature of CMT systems in the study area. Next, we consider the issue of whether CMT systems are sustainable. Finally, we conclude with the implications for government policy.

The term 'customary marine tenure' was first coined by Hviding (1989) when he used it to refer to particular forms of sea tenure practiced in the Pacific Islands.[27] Hviding uses the term '*customary*' to refer to a system founded on traditional roots and with links to the past; '*marine*' refers to the fact that the system deals with coral reefs, lagoons, coast and open sea, including islands and islets; and '*tenure*' refers to the fact that the system deals with access to marine areas and regulation of exploitation. Although CMT systems are widespread in Pacific Island countries, including Australia, there have been few studies that have attempted to document or research them. In recent years, the governments of PNG, the Solomon Islands, Vanuatu and Fiji have formulated policies, which explicitly recognise 'tradition' and 'custom' in the process of economic development. For example, Section 5 of the Customs Recognition Act of the PNG

---

[26] This section draws heavily on Asafu-Adjaye (2000).

[27] See, for example, studies by Hviding (1989), Ruddle *et al.* (1992) and Hviding *et al.* (1994).

constitution confers rights over water, reefs, and seabed to traditional owners. The section states:

> *the ownership by custom of rights in, over*
> *or in connection with the sea or a reef; or in*
> *or on the bed of the sea or of a river or lake,*
> *including rights of fishing; or the ownership*
> *by custom of water, or of rights in, over or*
> *to water.*

Thus, PNG law acknowledges traditional rights of ownership to coastal water and fisheries resources. However, in reality, the government only recognises CMT when it is practised within three miles of the coastline. The problem is that such boundaries are often in conflict with those of CMT systems. CMT systems use the estuaries of rivers, seashores and mangrove swamps as landward reference points. From these points, the sea territory extends outwards to include ocean-facing submerged reefs[28].

Coastal communities have connections with their marine environments that extend beyond three miles of the coastline. These areas are referred to as 'home reefs'. Communities often have folklore that asserts that their ancestors were the original inhabitants and users of these particular areas. Indigenous coastal communities have a holistic connection to the sea and certain marine sites are of special religious and cultural significance. For these and other reasons, some CMT systems may not have well demarcated boundaries. For example, some species of socio-cultural significance (e.g. dugong and turtle) are migratory and regarded as an inseparable component of the seascape. Therefore, the community's connections to these animals and the marine environment may transcend government-imposed boundaries.

A major issue, which has been the subject of some debate, is whether traditional resource management systems, of which CMT systems are a part, are inherently sustainable. A popular but mistaken view among many resource managers and government policy-makers is that traditional resource management practices today are no longer sustainable because modern economic forces have undermined them. One of the arguments that have been used to support this view is that indigenous societies are no longer in

---

[28] In one Solomon Islands coastal community, the marine boundary is defined as the farthest point one can see from the tallest coconut tree.

balance with their environment since they now use upgraded technology such as outboard motors and guns rather than canoes and bows and arrows. Such views and misconceptions are based on lack of knowledge or ignorance about how indigenous communities interact with their natural environments. Another common misconception is that poor or marginalised communities do not hold strong environmental values because they are only concerned about survival. However, the fact of the matter is that tradition is not static.

CMT systems, like the societies from which they come, are dynamic and highly adaptive. Over time, CMT systems in the study area have changed to reflect the changing social and economic circumstances. The evidence suggests that CMT systems thrive on an impressive body of local environmental knowledge amassed over a long period of time. This knowledge also includes an awareness of ecological interaction between land and sea. Decisions about resource use are made on the basis of such knowledge and there is little doubt that, in terms of resource management in local areas, CMT systems are capable of sustaining marine resources.

Obviously, as economic conditions worsen, CMT systems are put under pressure. In the study area, there is evidence of increasing commercialisation of certain common property resources. Also, the dire economic circumstances of the people have caused some to abandon or overlook certain CMT rules. However, these developments do not necessarily mean that CMT systems have failed and must be abandoned. On the contrary, the government must make efforts to strengthen them. CMT systems are communal, exclusive access collective property systems. They are self-managed and self-regulated systems that change according to local circumstances. The World Conservation Strategy recommended CMT systems as one way of enhancing common property resources (IUCN, 1990:33), and they were also given a prominent role in Agenda 21 and the Rio Declaration (UN, 1992).

CMT systems meet the requirements for social equity since all members of the community are given input in decision-making. In particular, CMT systems support multi-purpose and diversified exploitation strategies. This is in line with the fact that most village households operate mixed production strategies. CMT systems involve not only controlling access to resources, but also policing and monitoring. It may be argued that they constitute a sustainable management tool.

Customary marine tenure systems have a potential role to play in the sustainable management of fisheries and marine resources in Papua New

Guinea, as well as in other PICs. In the past, fisheries managers and government policy-makers have wrongly applied the fisheries management model advanced by Hardin's (1968) 'tragedy of the commons'. According to this model, where access to a fishery is free, it is not in the interest of anyone to limit his or her effort. To prevent overfishing and eventual destruction of fish stocks, fisheries managers have tended to impose rules based on limits on fishing effort. Such methods have not been successful as management tools because they have been imposed on the community by outsiders. CMT systems are more likely to succeed in the management of local fisheries resources because they are community-based and are derived from kinship and lineage structures. It is advocated that government and fisheries resource managers must recognise and strengthen the functioning of CMT systems to enable them play their role in the sustainable management of fisheries resources.

References

Asafu-Adjaye, J. (2000). Customary Marine Tenure Systems and Sustainable Fisheries Management in Papua New Guinea: The Case of the Western Province. *International Journal of Social Economics* 27 (7/8/9/10):917-926.

Hardin, G. (1968). The Tragedy of the Commons. *Science,* 168: 1243-1248.

Hviding, E. (1989). *All Things in Our Sea: The Dynamics of Customary Marine Tenure, Marovo Lagoon, Solomon Islands.* Special Publication No. 13, National Research Institute, Boroko, PNG.

Hviding, E., Baines, G.B.K and Ghai, D. (1994). *Community-Based Fisheries Management, Tradition and the Challenges of Development in Marovo, Solomon Island.* Blackwell Publishers, Oxford, U.K

IUCN-UNEP-WWF (1990). *Caring for the World: A Strategy for Sustainabiliy.* Second Draft, World Conservation Union, Glands, Switzerland.

United Nations, UN (1992). *Rio Declaration,* Final text of agreements negotiated by governments at the United Nations Conference on Environment and Development, 3-14 June, Rio de Janeiro, Brazil, Principles 7-8, 11-12, 14-15.

Part II. Tools for Environmental Policy Analysis

5.   Environmental Valuation

Objectives

After studying this chapter you should be in a position to:

□   explain the various types of non-market values associated with environmental goods and services;

□   explain various techniques for estimating non-market values;

□   explain the limitations of these techniques; and

□   implement the travel cost method.

5.1   Introduction

The proper valuation of non-market environmental commodities has significant policy implications. In the past such commodities have been assigned zero or low values due to difficulties involved in assigning economic values. Failure to properly account for the values of some environmental resources has resulted in decisions that have had negative implications for the environment and the society. Environmental valuation is also important in the event of natural disasters, either man-made or naturally occurring. A recent example is the Exxon Valdez oil spill in Alaska. In a landmark case the United States Supreme Court ruled that the people of Alaska should be compensated for their loss of livelihood and recreation as a result of the accident. As most of the lost benefits were non-market in nature, conventional market techniques could not be used.

This chapter begins with a description of the types of non-market values associated with a given environmental resource. The chapter then proceeds to describe a range of techniques for valuing non-market environmental commodities. A summary of the main points is provided at the end.

5.2   Types of Economic Values

Whenever we think of a resource, say a forest, the value that immediately comes to mind is the utility that we derive from direct uses (e.g., timber harvesting, and recreation). However, there is a range of values that is associated with a natural resource such as a forest. The **total economic value** (TEV) of a natural resource can be divided into two broad categories: **instrumental** or **use value**, and **intrinsic** or **non-use** (or **passive use**) **value** (Figure 5.1). Use values, which are most commonly known, refer to the capacity of a good or service to satisfy our needs or preferences. Use values can be further divided into **direct value** and **indirect use value**.

Figure 5.1 A taxonomy of economic values

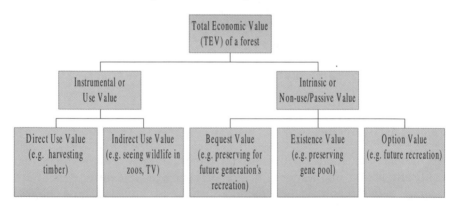

Direct use values consist of **consumptive uses** such as timber harvesting and **non-consumptive uses** such as camping, hiking and birdwatching. Indirect use values include environmental services such as maintenance of the hydrological system, climatic stabilization (e.g., carbon fixing) and soil stabilization.

Intrinsic or non-use values, as the name suggests, are inherent in the good. That is, the satisfaction we derive from the good is not related to its consumption, *per se*. Non-use or passive use values comprise **existence value, bequest value** and **option value.**

Existence value arises from the benefit an individual derives from knowing that a resource exists or will continue to exist regardless of the fact that they have never seen or used the resource, or intend to see or use it in the future. A good example of the significance of non-use value is the international outcry over the whaling issue. There are many people who have never seen a whale or plan to see one, but are nevertheless willing to pay significant sums of money to ensure that whales are not hunted to extinction.

Bequest values arise from the benefits that individuals derive from knowing that a resource will be available for their children and children's children. Option value is a little more complex. Option value may be defined as the amount of money an individual is willing to pay, at the current time, to ensure that a resource is available in the future, should they decide to use it. To the extent that option value is the expected value of future use of the resource, it may also be classified as a form of use value.

A related type of option value is **quasi-option value** (Arrow and Fisher, 1974; Fisher and Hanemann, 1987). Suppose there is a choice between conservation and development. However, the development option will result in an irreversible change. In this case, quasi-option value is the value of information that results after a decision has been made to develop or conserve at the present time. For example, if a cure for a fatal disease were to be found after the conservation decision has been made, then quasi option value would clearly be positive. It must be noted that quasi-option value cannot be summed with option value because it measures a different concept.

Use values can be readily measured by market prices or other means and are well accounted for in decision-making processes. However, as indicated earlier, non-use values are problematic because they are not traded and therefore cannot be valued by market prices. Empirical research suggests that non-use values can be a significant component of total economic value. Table 5.1 reports estimates of use and non-use values for wildlife in Alberta, Canada. The preservation (i.e., non-use) benefits of wildlife were estimated to be C\$67.7 million per annum out of a total economic value of C\$185.2 million per annum in 1987 dollars. In this case non-use benefits were at least one-third of TEV. Thus, failure to consider such benefits, whether quantitatively or qualitatively, in the decision-making calculus could lead to gross underestimation of the contribution of wildlife to total social welfare.

We discuss below a range of techniques for obtaining non-market values of environmental resources.

Table 5.1 Estimates of the economic value of wildlife in Alberta, Canada

Type of value	Mean values C$/capita/p.a.	Number of participants	Annual values (1987 C$mil.)	In perpe-tuity (C$ mil. p.a.)
Preservation Benefits	80.9	836,125	67.7	1354
Hunting				
Waterfowl	171.8	59,730	10.3	206
Other birds	130.0	84,827	11.0	220
Small mammals	119.1	56,738	6.8	136
Large mammals	211.1	118,207	24.9	498
All hunting	165.9		53.0	1060
Non-Consumptive Use	163.0	395,873	64.5	1290
Total Economic Value			185.2	3704

Source: Adamowicz, Asafu-Adjaye, Boxall and Phillips (1991).

5.3   Non-Market Valuation Methods

Non-market valuation methods can be broadly classified into two categories: **revealed preference (RP)** models, **stated** (or **expressed) preference (SP)** models, and combined SP and RP models. Revealed preference approaches make use of individuals' behaviour in actual or simulated markets to infer the value of an environmental good or service. For example, the value of a wilderness area may be inferred by expenditures that recreationists incur to travel to the area. The value of, say, noise pollution may be inferred by analysing the value of residential property near an airport. These methods are also referred to as indirect or surrogate market approaches.

Examples of RP methods include:
- Travel Cost Method (TCM)
- Hedonic Pricing Method (HPM)
- Cost (or Expenditure) Methods, and

- Benefit Transfer Methods

Stated preference methods attempt to elicit environmental values directly from respondents using survey techniques, hence the alternative name of 'direct approach'. As will be explained below, these methods are flexible and can be applied to a wider range of environmental goods and services than RP methods. Furthermore, SP methods can be used to estimate total economic value (i.e., use and non-use values), whereas RP methods can be used to estimate only use values. Stated Preference methods do have some drawbacks and these are discussed below.

5.3.1 Stated Preference Models

Stated preference models can be further classified into three types: contingent valuation method (CVM), conjoint analysis, and choice modelling. In this section we first consider the CVM and then go on to discuss conjoint analysis and choice modelling.

The Contingent Valuation Method

The CVM uses interview techniques to ask individuals to place values on environmental goods or services. The term 'contingent' in CVM suggests that it is contingent on simulating a hypothetical market for the good in question. The most common approach in the CVM is to ask individuals the maximum amount of money they are willing to pay to use or preserve the given good or service. Alternatively, the respondents could be asked the maximum amount of money they are **willing to accept in compensation** (WTA) to forgo the given environmental good or service. Theoretically, these two measures should be equivalent. However, empirical studies have indicated that WTA estimates exceed WTP estimates.[29] Typical steps in a CVM procedure are as follows:

1. Set up the hypothetical market;
2. Obtain the bids;
3. Estimate mean WTP and/or WTA; and

---

[29] Critics of the CVM assert that this is an indication of the lack of validity of the method. However, recent research indicates that the divergence should be expected on both economic and psychological grounds. Individuals may feel the cost of a loss more intensely than the benefit of a gain.

4.   Estimate bid curves

Setting up the hypothetical market

The first step is to establish a reason for a good or service where there is no current payment. Suppose there is a government proposal to mine, say, a wilderness area. Assuming few people actually visit the area, the analyst would describe the area and the impacts of the proposed project on the environment. Pictorial aids could also be used in setting up this hypothetical market (not applicable to a telephone interview).

Obtaining the bids

The second step is to decide on a suitable 'bid vehicle'. This is the method by which the WTP or WTA bids would be elicited. Possible bid vehicles could include income taxes, property taxes, utility bills, entry fees, and payments into a trust fund.

Methods used to obtain the bids include face-to-face interviews, telephone interviews or postal surveys. A face-to-face interview allows more scope in presenting the hypothetical market and clarifying respondent concerns. However, it is the most expensive method because interviewers have to be paid. The telephone interview and postal survey offer less flexibility, in declining order

Methods of obtaining bids include the following:

- **Bidding games**: respondents are offered progressively higher bids until they reach their maximum WTP.
- **Payment card**: a range of values is provided on a card and the respondent is requested to choose one.
- **Open-ended questions**: respondents are asked to report their maximum WTP.
- **Close ended questions**: there are at least three variants:
    - (i) **Dichotomous choice (referendum)**: a single amount is offered and respondents are asked to provide a 'Yes' or 'No' answer, also referred to as the 'take it or leave it' approach;
    - (ii) **Double-bounded referendum**: respondents who answer 'no' to the first amount are offered a lower amount, and those who answer 'yes' are offered a higher amount; and
    - (iii) **Trichotomous choice**: respondents are offered three choices to the payment—'yes', 'no' and 'indifferent'.

An important requirement of using the CVM is that the respondents must be reminded of their budget constraints when eliciting their bids. The dichotomous choice (or referendum) format (Bishop and Heberlein, 1979) is considered to be the state-of-the-art in CVM methodology. A National Oceanic and Atmospheric Administration (NOAA) panel of economic experts, chaired by Kenneth Arrow and Robert Solow, recommended the referendum format over the open-ended format (NOAA, 1993). The double-bounded referendum (Hanneman, 1985) and trichotomous choice (Ready *et al.*, 1995) are relatively more recent variants of this approach.

Estimating mean WTP and/or WTA

For the first three bid elicitation approaches the mean and median WTP can be found from the individual bids. Mean and median bids for the close-ended referendum bids are more difficult to obtain. Analytical methods such as probit, logit and random utility models can be used to obtain estimates.

Estimating the bid (demand) curves

Bid (or demand) curves could be estimated at this stage to validate the WTP results and to estimate aggregate WTP. The bid curve is estimated by regressing WTP against relevant socioeconomic variables, and checking to see whether the signs conform to theory. For example, the following demand function could be estimated:

$$WTP_i = f(A_i, E_i, Y_i, M_i) \tag{5.1}$$

where A is age; E is educational level; Y is income level and M is a variable for membership of an environmental organisation. Based on economic theory, we would expect Y to be positively related to WTP.

The total value of the good or service can be estimated by multiplying the mean WTP by the number of households (if the sampling unit used was the household).

Choice Experiments

Choice experiment approaches include **conjoint analysis**[30] and **choice modelling** (CM). Conjoint analysis is further divided into **contingent ranking**, **contingent rating** and **paired comparison**.

Conjoint analysis

A major difference between CVM and conjoint analysis is that in the former respondents are required to evaluate only one or sometimes two alternatives. On the other hand, the latter requires them to evaluate several alternatives separately. In contingent rating, respondents are requested to rate their preferences for several alternatives on, say, a ten-point scale. They are presented with a set of attributes associated with each alternative. The respondents' ratings are then regressed against the attributes. The marginal rate of substitution between a given attribute and its price provides an estimate of the 'value' of the attribute. This is referred to as the **'part-worth'** of the attribute. Summing all the part-worths provides an estimate of a respondent's WTP for an aggregate change in the environmental good or service. In contingent ranking, respondents are required to rank all the alternatives from least preferred to most preferred. The analysis of contingent ranking data is similar to that of contingent rating. The rankings can be converted to a ratings scale and analysed with multiple regression techniques, or other estimation methods such as logit or probit analysis can be used.

A weakness of both contingent rating and contingent ranking is that they do not provide the respondent with an opportunity to reject the good. The only way they allow opposition is by registering a low rating or ranking. In that sense these methods are considered to be unconditional or relative measures of WTP and could be understated.

In the paired comparison approach, respondents are presented with successive sets of two choices and asked to rate the difference between them on a scale (usually, a 5-point scale). A form of paired comparison is adaptive conjoint analysis where the pairs are generated with the aid of a computer. Like the previous two methods, the data from the paired comparison can be analysed using **multiple regression**, **logit** or **probit models** to provide

---

[30] Conjoint analysis is a popular technique in marketing research. It has only recently been adapted for valuing environmental goods and services.

estimates of a respondent's WTP for an aggregate change in the environmental good or service.

Choice modelling

Choice modelling was developed by Jordan Louviere and was initially used in the field of marketing to analyse consumer choices (Louviere and Woodworth, 1982). Since then, a few studies have used the method to value environmental goods and services.[31] In this approach, respondents are presented with a series of alternatives, with each containing three or more resource use options. Usually, each alternative is defined by a number of attributes. For example, in a CM study of preserving a wilderness area the attributes could be the following: numbers of rare species present; ease of access to the area, size of area and cost to households (Box 5.1). These attributes would then be varied across the various alternatives. The respondents are then required to choose their most preferred alternative. Estimates of respondents' WTP are obtained by estimating a multinomial logit model.

Choice modelling is relatively more versatile than the other SP methods. It can be used to value multiple sites or multiple use alternatives. Unlike conjoint analysis, CM can be used to provide conditional or absolute measures of WTP provided a 'choose neither' option is included among the alternatives.

The main disadvantage of choice modelling is that complex survey designs are required. The number of choice sets can be large, which tends to lengthen interview times.

Biases associated with stated preference models

Stated preference models have some distinct advantages. Most are straightforward to apply and do not require any theoretical assumptions compared to revealed preference approaches. The only assumption is that the individual is able to value the good or service and will truthfully report his or her valuation. Furthermore, at the present time, methods such as CVM and choice modelling are the only ones that can be used to estimate non-use or passive use values.

---

[31] See applications by Adamowicz *et al.* (1994) for valuing water-based recreation and by Morrison *et al.* (1998) for valuing wetlands.

Box 5.1 Choice modelling application in the Desert Uplands,
Central Queensland

Choice modelling was used to estimate the economic value of preventing the loss of endangered species in the Desert Uplands region of central Queensland (Blamey *et al.*, 1997). The aims of the study were: (i) to apply and develop the choice modelling technique; and (ii) to provide estimates of the economic value of preventing endangered species loss to assist resource management decision-making. The first objective was to enable a test of choice modelling as an alternative to the CVM given the problems associated with the latter. With regards to the second objective, land clearing is a major problem in the area, and there was a need to assess the trade-offs involved with alternative land uses.

The preliminary results suggest that the loss of endangered species was valued at A$11 per species by Brisbane residents aged under 30 years of age and $14 per species by Brisbane residents aged under 60 years of age.

Source: Blamey *et al.* (1997).

In spite of (or maybe due to) their simplicity, stated preference methods are subject to a number of biases. These include: (i) **hypothetical bias**; (ii) **embedding effect**; (iii) **strategic bias**; (iv) **bid vehicle bias**; (v) **starting point bias**; (vi) **information bias**; (vii) **part-whole bias**; and (viii) **non-response bias**.

These biases are briefly discussed below.

(1) Hypothetical bias: the major assumption in the CVM is that the amount of money people say they are willing to pay corresponds to the individual valuations of the good or service in question. The CVM has been criticised for the fact that respondents do not actually have to pay their stated amounts. Therefore, it has been suggested that the hypothetical nature of the exercise might induce people to 'free ride', that is, understate their true WTP. However, in a series of experiments in which hypothetical WTP has been compared to actual WTP, hypothetical bias has not been found to be significant.

(2) Embedding effect: embedding effect (Kahneman and Knetsch, 1992) is said to occur when an individual's WTP is lower when it is valued as part of a more inclusive good or service, rather than on its own. It has been suggested that embedding effect occurs because people are seeking a 'feel good' or 'warm glow' associated with contributing to a 'good' cause. Some researchers attribute embedding to the existence of substitutes. That is, people will reduce their WTP if they are aware of substitutes. Embedding effect is minimised in CM because it allows explicit inclusion of substitutes.

(3) Strategic bias: strategic bias occurs when a person deliberately overstates (or understates) his or her true bid in order to influence the outcome. For example, some people who strongly support a proposed development may report a zero WTP for conservation even when they have a positive WTP. Other SP methods may not suffer from the same level of strategic bias as CVM because they do not require respondents to state their bids.

(4) Bid vehicle bias: as noted above, the CVM depends on a 'vehicle', that is, a means by which the stated hypothetical amounts would be collected. An individual who dislikes a particular kind of vehicle (e.g., higher taxes) may understate his or her WTP. In some areas, respondents might be dissatisfied with the way their government is using their taxes and therefore such a vehicle might invoke a negative response. A solution to this problem is to use a 'neutral' vehicle. For example, for preservation values, a useful vehicle could be donations to a trust fund to be administered by an independent non-governmental organisation. Vehicle bias is present in CVM and contingent ranking. There is a question mark on the presence of vehicle bias in contingent rating, paired comparison and CM because these methods emphasise multiple attributes which place less emphasis on payment.

(5) Starting point bias: some CVM bid elicitation formats (payment card and bidding game) 'start' off with a certain amount. The stated amount may induce bias in the sense that it may be misinterpreted by the respondent as a cue for an 'appropriate' range of WTP. There is not much empirical evidence about the extent of this type of bias. However, it may be minimised by extensive pretests of the survey instrument.

(6) Information bias: because a CVM is conducted by creating hypothetical scenarios, this scenario must be conveyed to the respondent by providing information. The quantity, quality and sequencing of this information can influence the bids. Insufficient information will make it difficult for the respondent to properly value the given good if he or she has no prior knowledge of it. On the other hand, too much information would be a definite source of bias. One way of minimising this kind of bias is to provide enough information to model the real context of the valuation exercise.

(7) Part-whole bias: there is concern that if people are asked to value one part of a given asset (e.g., all wildlife) and then subsequently asked to value a part of it (e.g., a given species) the response may be similar. It has been suggested that this problem arises from the way people allocate their personal budget, first dividing their income amongst broad consumption categories, and then allocating to sub-categories of goods. The solution to this problem is to remind them of their budget constraints and to restrict valuation to whole goods rather than parts of the good.

(8) Non-response bias: this type of bias is associated with surveys, in general. Some people cannot be bothered to participate in surveys. Often, it is those with particular interests in the subject who are likely to respond. In such cases, it may be argued that the sample is not representative of the population. Non-response bias can be minimised if questions are easier to answer. In this regard, it has been suggested that the CM format may be easier than, say, the CVM. Mackenzie (1993) reported that only 1.4 percent of respondents in a CM survey refused to answer rating questions.

Summary of stated preference methods

Stated preference methods are relatively straightforward approaches for eliciting individuals' valuations of non-market environmental goods and services. They require few theoretical assumptions. For example, the only assumption implicit in the use of the CVM is that respondents have an idea of their personal preferences and are willing to truthfully report their willingness-to-pay. However, the validity and reliability of estimates obtained using these methods may be questioned due to inherent biases.

Most of these biases are associated with the CVM, in particular. These include hypothetical, part-whole, strategic, vehicle, starting point and non-response bias. The other SP approaches are less prone to certain types of biases. Recent research, however, suggests that following certain best practice procedures in survey design can minimise most of these biases. At the present time, CVM and CM are the only techniques that can be used to value non-use values.

5.3.2 Revealed Preference Models

Revealed Preference models include the travel cost method, the hedonic price method, market value or cost methods and the benefit transfer method

Travel Cost Method (TCM)

The concept of using travel costs to value recreation was first proposed by Hotelling in 1949 and formalised by Clawson (1959). However, Davis (1963) was the first to apply the method in a study of the value of recreation in Maine forests. The basic assumptions underlying the TCM are that (i) the costs an individual incurs in visiting a recreational site reflect the person's valuation of that site, and (ii) individuals will react to an increase in entry fees in the same way as they would react to an increase in travel costs. That is, at some high level of entry fee (or cost of travel) no one would visit the site because it would be too expensive. By asking visitors questions relating to where they have travelled from and the costs they have incurred, and relating this information to the number of visits they make per annum, a demand curve can be generated for the recreational site under question. This curve will be downward sloping in the sense that travel cost will be inversely related to number of visits (Figure 5.2). That is, those living near the site will make more visits per annum compared to those living far away.

There are two forms of the travel cost method: the **zonal travel cost method** (ZTCM) and the **individual travel cost method** (ITCM). In the zonal travel cost approach (e.g., see Smith and Kaoru, 1990), concentric zones are defined around each site such that the cost of travel from all points in a given zone is approximately constant. Visitors to the site are grouped according to their zone of origin. By comparing the cost of coming from a zone with the number of people who come from it and the population of the zone, one can plot a point for each zone. A curve can then be fitted to all the points to generate the demand curve from which a measure of consumers surplus can be obtained.

Figure 5.2 Demand curve for the travel cost method

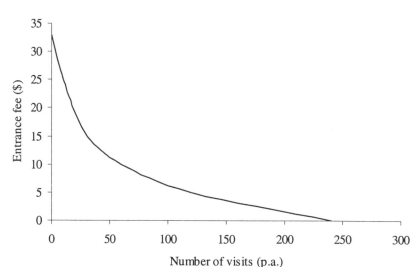

A trip-generating function for the ZTCM can be specified as:

$$V_h/N_h = f(C_h, X_h) \tag{5.2}$$

where: V_h = number of visits from zone h
N_h = population of zone h
C_h = travel cost from zone h
X_h = a vector of socioeconomic variables that explain changes in V

The ITCM uses the number of visits per annum made by an individual, rather than zonal visits, as the basis for generating the demand curve. The trip-generating function for the ITCM can be stated as follows:

$$V_i = f(C_i, X_i) \tag{5.3}$$

where: V_i = number of visits made by individual i to the site
C_i = cost of a visit by individual i to the site

X_i = socioeconomic factors affecting individual i's visits to the site

By integrating the area under the demand curve, an ITCM estimate of the individual's consumer surplus can be obtained. This figure is then multiplied by the number of visitors per annum to obtain the aggregate benefits.

Recently, variations of the travel cost model have been developed. These include the combined TCM and CVM model (discussed in more detail later), the random utility model (RUM), the **hedonic travel cost model**, and the **multi-site travel cost model**. The RUM (Ward and Loomis, 1986) first models the visitor's decision on whether or not to participate in the recreation activity, followed by the decisions on the number of visits. Econometric techniques such as probit, tobit and and logit models are used to estimate these two decisions independently. The hedonic travel cost model (Bockstael *et al.*, 1991) identifies separate characteristics of the recreation experience and the price that people are WTP for each. The first stage of the procedure involves regressing the respondents' travel costs on measures of quality characteristics at the site. A separate demand curve for each characteristic is then developed. The multi-site travel cost model (Ward and Beal, 2000) attempts to account for all relevant substitute sites within the region. Visitation data from multiple sites that have different levels of the site quality variable of interest are used and the model predicts demand for each site.

Table 5.2 reports travel cost estimates for various protected areas in Queensland. Beal (1995) estimated a value of A\$2.50 per person per visit for trips to the Carnarvon Gorge National Park, but Hundloe *et al.* (1990) estimated relatively higher values of A\$15.70–32.63 per visit to Fraser Island. Scoccimarro (1992) estimated the value of recreation in the Green Mountains, Lamington National Park, to range from A\$8.09–9.23 per visitor per day. Much higher estimates of A\$362 per visit and A\$49 per visit, respectively, for Hinchinbrook Island (Stoekl, 1994) and the Wet Tropics World Heritage Area (Driml, 1996) have been made.

Limitations of the TCM

As indicated above the main underlying assumption of the TCM is that the value of a recreational site corresponds to the costs that the respondent incurs in undertaking the recreational experience. A distinct advantage of the TCM is that it is based on real rather than hypothetical data and as such can provide true values. It is based on the simplified assumption that the recreational value of a place is directly related to travel costs incurred in getting there.

Table 5.2 Travel cost estimates of protected areas in Queensland

Site	Study	Average willingness-to-pay
Carnarvon Gorge National Park	Beal (1995)	A$2.50 per person
Fraser Island	Hundloe *et al.* (1990)	A$15.70–32.63 per visit
Green Mountains, Lamington National Park	Scoccimarro (1992)	A$8.09–9.23 per visitor day
Hinchinbrook Island	Stoekl (1994)	A$362 per visit
Wet Tropics World Heritage Area	Driml (1996)	A$49 per visit (domestic tourists)

However, the TCM suffers from the following limitations.

(1) The TCM is suited to estimating the value of particular sites or locations and is unsuited for measuring other kinds of goods or services. For example, TCM cannot be used to value non-use or passive use values. This is because it is based on the travel costs of users of a given recreational site. Non-users, who may have significant values for the same site are excluded from the sample.

(2) Multiple destinations: a problem arises about the appropriate allocation of costs among multipurpose journeys. The allocation of such costs could be arbitrary and there is currently no consensus on how to do this. A variety of ways of dealing with this issue has been suggested in the literature. One approach is to omit multiple-destination visitors and consider only single-site visitors. The other is to ask the respondents to allocate a proportion of the total travel cost to the given site.

(3) Visits to certain sites could be seasonal and therefore the survey results could be biased unless it is conducted over a long period.

(4) Travel costs: the assumption that travel costs reflect recreational value may not always be true. For example, people who live near the site may

incur zero or minimal travel costs but may nevertheless have high values.

(5) Time and other factors: the TCM assumes that travel costs (e.g., fuel costs) are the major determinants of the value of a recreational site. However, other factors could affect the demand for recreation. For example, travel time is an opportunity cost because the time spent travelling is not available for other pursuits. Time should therefore be considered as a cost.[32] However, there is no consensus as to how time should be accounted for in TCM. In some studies, a certain proportion of the wage rate is multiplied by travel time to provide an estimate of the opportunity cost of time. However, the choice of the weight is quite arbitrary and open to question.

(6) Welfare estimates computed from the TCM may differ depending on the variables and functional form used to estimate the relationship between demand for visits and the cost of travel.[33]

Summary of TCM

To summarise, the travel cost method is a useful method for valuing the recreational benefits of a site. However, it is a restrictive method in the sense that it can only be used to measure site-specific recreational value. There are also some problems in actually deriving benefit estimates. These relate to the issues of the cost of travel time and substitute sites.

The Hedonic Price Method

The **hedonic price method** (HPM) derives values for an environmental good or service by using information from the market price of close substitutes. It is based on Lancaster's consumption theory which assumes that a good or service provides a bundle of characteristics or attributes (Lancaster, 1966). Suppose the government wishes to value the disutility generated by aircraft noise in a given location. It could do so by analysing variations in house prices as one moves away from the flight path of aircraft. Take the example of two houses with the same facilities (e.g., number of

---

[32] If a person enjoys, say, views of the countryside while travelling, or simply enjoys travelling then, of course, this should be considered a benefit.

[33] This problem also applies to the hedonic price method.

bedrooms, bathrooms, and swimming pool). One is directly under the flight path and the other is quite a distance away. It is expected that the house under the flight path will be cheaper and the price difference may be attributable to the value of the noise pollution.

Hedonic price models attempt to explain individuals' willingness-to-pay in terms of a set of attributes and characteristics of the good. For example, the price a potential homebuyer is willing to pay for a house depends on the location of the property, number of rooms, access to amenities (e.g., schools and transport), the environmental quality and the person's income. The hedonic price equation may therefore be expressed as

$$\text{Price of house} = f(\text{location, number of rooms, access to amenities, income of buyer, environmental quality}) \quad (5.4)$$

where environmental quality is proxied by aircraft noise measured in decibels.

Data are collected on each of the five variables for a reasonable sample of houses. Equation (5.4) is then estimated using multiple regression techniques. We would expect house prices to be positively related to the number of rooms, positively related to the degree of access to amenities (e.g., shops, schools, entertainment), positively related to income and negatively related to environmental quality (aircraft noise in decibels). The monetary value for a one-unit change in noise level can be found by differentiating this function with respect to environmental quality, or alternatively, by finding it from a plot of the function (Figure 5.3).

Table 5.3 presents HPM estimates for the value of traffic noise in the U.S. The results reported represent coefficients of the hedonic price function for various American cities. They measure the percentage fall in house prices due to a one-decibel increase in noise levels. The monetary value of noise is the coefficient multiplied by the average house price in the area. For example,

Monetary value for a one decibel decrease in noise in North Virginia

$$= 0.15 \times (\text{average house price in North Virginia})/100 \quad (5.5)$$

Figure 5.3 Effect of aircraft noise on house prices

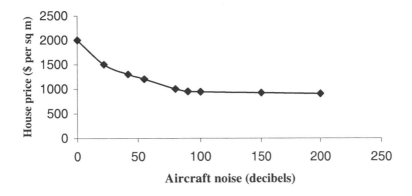

Table 5.3 Impact of traffic noise on house prices in the U.S.

Area of U.S.	Percent fall in house price due to a 1 decibel increase in noise level
North Virginia	0.15
Tidewater	0.14
North Springfield	0.18–0.50
Tourism	0.54
Washington, D.C.	0.88
Kingsgate	0.48
North King County	0.40
Spokane	0.08
Chicago	0.65

Source: Nelson (1982).

Summary of the HPM

The HPM is suited to the estimation of the characteristics of goods and services. However, it also has some limitations. As is the case for the TCM, welfare estimates computed from the HPM are sensitive to the choice of functional form and the variables included in the model. The HPM is susceptible to the problem of multicollinearity, which refers to a high degree of correlation amongst the explanatory variables in the hedonic price function.

A major assumption is that, given income constraints, people are free to select the characteristics of houses that satisfy their preferences and that the price they are willing to pay takes account of these factors. However, house prices can also be affected by external factors such as taxes and interest rates, which are not accounted for in the hedonic price equation.

Market Value Method

Market value approaches make use of observed market prices for environmental goods and services. Based on our classification of TEV above, it can be seen that this approach can only be used to value environmental goods and services that have established markets. These are commodities which have:

- direct uses: e.g., plantation timber, commercial fisheries, tourism;
- some indirect uses: e.g., the value of water from protected watersheds; and
- some option values: e.g., gene research, forest conservation.

Market value methods attempt to find a link between a proposed environmental change and the market value of the corresponding goods and services. A common approach is to use changes in productivity of the good or service. For example, the direct impacts of an environmental change on human health can be estimated as a change in income. The assumption here is that sickness reduces one's ability to earn income.

The advantages of the market value method are:

(1) It is relatively simple and straightforward;
(2) It relies on actual market values; and
(3) It has some relation to measured output.

The disadvantages are:

(1) It is limited in the types of values that it can capture;
(2) It can be difficult to define the physical flows over time;
(3) In some cases, the links between the environmental change and the market good or service may not be obvious.

Market Cost Method

In general, market-cost methods measure the cost of achieving a particular objective. Examples include restoring certain environmental services and avoiding land degradation. These methods focus on the cost of prevention or rectifying environmental damage and the cost of replacing environmental services. Most often, these costs are estimated from market prices, including the costs of labour and materials used in the particular activity.

There are a number of variations of the market cost method. These include the following: (1) change in cost; (2) replacement cost; and (3) defensive expenditures. The basic assumption of cost methods is that the value of a good is equal to some multiple of the cost of producing it.

(1) **Change in cost method:** suppose a proposed project may change the cost of a good or service. If the project causes a decrease in the cost of the good or service, this can be interpreted as a gain in benefits, that is, a cost saving. On the other hand if the project results in an increase in costs, this may be taken to be a loss of benefits. Take the example of a project that involves the construction of a water supply system. In this case a major benefit is the cost savings to households from not having to buy water from water vendors or transport water over long distances.

(2) **Replacement cost method:** assumes that the value of an existing good or service is the cost of replacing it. For example, if a storm damages roads, buildings and transmission lines then an estimate of the damage done is the cost of replacing these structures. However, in this case, the replacement cost must be considered as the minimum value of the benefits derived from these goods. This is because we need to add on the consumer surplus that people derive from utilising the good or service. A variation of replacement cost is **mitigation cost**. Mitigation cost is an estimate of the cost of restoring a damaged environmental good to its former condition. This approach could be useful where the damage is minor. Obviously, it is of limited use where the damage is either irreversible or total restoration is impossible.

(3) **Defensive-expenditure method:** in this approach the net benefit of a particular project intervention is the amount of money people are willing

to spend to either mitigate or avoid the impacts.[27] A good example is a community which does not have good drinking water. In this case, the benefits of introducing a water treatment plant include the amount of money people spend to boil or treat their water for cooking or drinking purposes.

Limitations of market and cost methods

The market and cost methods are easy to apply and can provide useful measures of net benefits. People can easily understand the use of monetary units. However, a major limitation is that they do not measure benefits that are determined from the interaction between the demand for and supply of environmental goods. As such, they only capture a portion of total benefits (e.g., they exclude non-use benefits). Where there is a high degree of non-market benefits or costs, market values may provide only minimum estimates of opportunity costs or forgone benefits.

Benefit Transfer Method

The benefit transfer method is another alternative for obtaining non-market values. This approach has been applied to value the impact of improved water quality on recreation values and public health (Kask and Shogren, 1994) and to lake recreation (Parsons and Kealy, 1994). It involves 'transferring' values that have been estimated for a similar good or service from another location to the current location. The approach is useful because surveys are expensive and, in addition to money, there could be a time constraint.

Economists are divided on the validity of the benefit transfer method. For this method to be meaningful, the following conditions must hold:

- The goods (or services) in both sites should have roughly similar characteristics;
- The population in both areas should be similar; and
- The values in the first study should not have been estimated a long time ago because preferences change over time.

Three tests have been suggested to determine the accuracy of benefit transfer (Box 5.1). The aims of these tests are to determine the convergent validity

---

[34] See Hufschmidt *et al.* (1993) for a good discussion of applications in developing countries.

(i.e., statistical validity) of benefit transfer and the extent of any bias. The first test involves comparing the benefit transfer values with primary data values obtained from the policy site. If the benefit transfer estimates are not statistically different from the primary data value estimates from the policy site, then it may be concluded that the benefit transfer values are valid. The extent of bias is given by the deviation between the two estimates.

The second test involves determining whether different populations have the same preferences for the same non-market good, after controlling for differences in socioeconomic characteristics such as income and education levels. The third type of benefit transfer test is to determine whether transfers are stable over time.[35] Many studies have concluded that value estimates remain relatively stable over a few years.

Morrison *et al.* (1998) investigated the suitability of using choice modelling estimates for benefit transfers both across different populations and across different wetlands in northern New South Wales, Australia. In general, the weight of the evidence appeared to be against the convergent validity of both transfers across sites and across populations. However, they found that transfers across sites tended to be less problematic compared to transfers across population.

5.3.3 *Combined Stated and Revealed Preference Models*

In the past, SP and RP models have been viewed as substitute valuation methodologies. Some researchers (e.g., Carson *et al.* 1996) have attempted to test the validity of one of the approaches by making comparisons of welfare estimates derived from both models. However, since each approach is subject to criticism, including a variety of biases and statistical estimation problems, it is not clear that such a validation strategy is effective. Cameron (1992) made the innovative suggestion that, rather than treating, say, the CVM and TCM as competing methods, the two approaches could be successfully combined to estimate welfare measures. Since information from two sources are being combined to estimate a given set of parameters, the combined model should be estimated more precisely than separate models.

Catherine Kling analysed the gains from combining TCM and CVM data using simulation experiments (Kling, 1997). She found that there were definite gains in precision from combining models and there also appeared to be gains in reduced bias. The parameter estimates from the combined model

---

[35] See, for example, studies by Reiling *et. al* (1990) and Teisl *et. al* (1994).

results are generally more efficient because information on the same set of underlying preferences is used to construct the estimates.

The advantages of combing RP and Sp methods can be summarised as follows:

(1) With a combined approach, researchers can afford to work with smaller samples since each person in the sample generates more than one observation;
(2) The combined approach results in improved statistical efficiency as indicated above; and
(3) The combined approach allows us to test for the consistency of SP and RP representation of individuals' preferences.

There are, however, some disadvantages of the combined approach. These are:

(1) The combined models are statistically complex and harder to implement;
(2) The combined approach does not work in all situations. For example, there must be a RP technique that fits the problem;
(3) In practical terms, there is limited experience in using this approach to analyse environmental issues; and
(4) Using the combined approach implies asking more survey questions. This could reduce the response rates, increase protests, and reduce the overall quality of the responses.

5.4   Summary

In this chapter various methods for estimating non-market values were introduced. The methods discussed were the SP methods (contingent valuation method, conjoint analysis, and choice modelling), RP methods (travel cost method, hedonic price method, the market value method and the market cost method), as well as combined SP and RP methods.

There is no single technique that is superior to the others. The choice of a particular technique depends on the particular resource valuation problem at hand. For example, if one wanted to estimate non-use or passive benefits, the CVM (or choice modelling) would be the technique of choice. If one wanted to estimate the recreation benefits of a particular resource, say, a national park, TCM would be a suitable choice. It was stated that there are gains in precision from combining SP and RP models and there also appeared to be gains in reduced bias. However, the combined models are complex and difficult to implement. Furthermore there is limited experience in the use of such models.

The application of these valuation techniques is important for decision making insofar as they take into account the unpriced or underpriced outcomes of proposed policies or projects. Although these techniques are not perfect, the inclusion of non-market values in the decision-making calculus helps to clarify the trade-offs and allows the decision makers to make better informed policy choices.

Key Terms and Concepts

benefit transfer method
bequest value
bid vehicle bias
bidding games
change in cost
choice experiment
choice modelling
combined revealed preference and
stated preference
consumptive uses
contingent ranking
contingent rating
contingent valuation method
dichotomous choice (referendum)
direct value
double-bounded referendum
option value
paired comparison
part-whole bias
passive use
payment card
probit model

embedding effect
existence value
hedonic price method
hedonic travel cost model
hypothetical bias
indirect use value
information bias
instrumental (use) value,
intrinsic (non-user) value
logit model
market cost method
mitigation cost
multi-site travel cost model
non-consumptive uses
non-response bias
open-ended question
revealed preference
starting point bias
stated (or expressed) preference
strategic bias
total economic value
travel cost model

quasi-option value   trichotomous choice
random utility model   willing to accept in compensation
replacement cost

Review Questions

1. List the different kinds of values associated with a marsh.

2. List some of the possible values associated with soil conservation.

3. Discuss the main differences between revealed preference and stated preference methods.

4. List and explain the biases associated with the stated preference approach.

5. Select a natural environment in your area. Develop a willingness-to-pay question to estimate the value of this natural environment. Suggest a way to check the validity of the responses.

6. Compare and contrast the contingent valuation method and choice modelling.

7. State any advantages and disadvantages of cost methods.

Exercises

1. The following table provides estimates of average house prices in the northern and southern areas of a city before and after a tollway was constructed near the northside.

Area	House price ($'000)		Number of houses
	Before	After	
Southside	150	210	15,000
Northside	150	100	10,000

Calculate the following:

    a.   Determine the value of houses in both locations after construction of the tollway.

    b.   Overall, is there a net gain/loss? How much is it?

2. A survey of 200 visitors to a national park gave the following results:

Total visit cost ($)	Number of visits per person per annum
140	0
120	1
80	3
60	4
30	6
20	7
16	10

    a.   Draw a demand curve for visits to the park as a function of the price, i.e., the travel cost.

    b.   The survey results indicate that 50,000 people living in the area surrounding the park take, on average, five visits to the park a year. Calculate the consumer surplus for a single visit to the park for the average person.

    c.   Calculate the total consumer surplus for an average visit.

    d.   Calculate the aggregate consumer surplus per annum for the national park.

3. The construction of a new irrigation system in a town is expected to benefit small-scale farmers living in and around the area. It is hypothesised that the development will lead to an increase in farm prices in the area. However, to benefit, a farmer must pay for the cost of constructing a pipeline to his or her property. A hedonic price model was estimated as follows:

$$\text{Farm price} = 500000 + 30\text{AREA} - 100\text{DIST} + 2000\text{IRIG}$$

where:

        AREA = farm size (ha);

        DIST = distance (in km) from the main distribution channel; and

IRRIG = dummy variable, where 1 = construction of irrigation scheme goes ahead; 0 = construction does not go ahead.

Given that the average farm size is 200 ha and is 15 km from the main distribution channel, calculate the following:

    a.   The average farm price if the project does not go ahead.
    b.   The average farm price if the project goes ahead.
    c.   The change in average farm price if the project goes ahead.

4. Data collected for possible admission fees to a zoo and corresponding number of visits are as follows:

Admission fee ($)	Total number of visits per day
0	8,000
1	6,000
2	4,000
3	2,000
4	0

    a.   Calculate the loss in consumer's surplus if entrance fees were to be increased from the current $1 to $3 per visit.

5. Suggest possible valuation methods for assessing the following:
    a.   Recreational fishing.
    b.   Water treatment and wastewater services.
    c.   Cyclone damage.
    d.   High voltage transmission lines.
    e.   Flood control programs.
    f.   Tourism and recreation
    g.   Loss of mangrove swamps.
    h.   Health effects of air pollution.
    i.   Creating artificial wetlands.
    j.   Stopping logging in World Heritage listed areas.
    k.   Improvement in waterway vegetation.

References

Adamowicz, W.L., Asafu-Adjaye, J., Boxall, P.C. and Phillips, W.E. (1991). Components of the Economic Value of Wildlife: An Alberta Case Study. *Canadian Field Naturalist*, 105(3): 423-429.

Adamowicz, W., Louviere, J. and Williams, M. (1994). Combining Revealed and Stated Preference Methods for Valuing Environmental Amenities. *Journal of Environmental Economics and Management*, 26: 271-292.

Arrow, K.J. and Fisher, A. (1974). Environmental Preservation, Uncertainty, and Irreversibility. *Quarterly Journal of Economics*, 88:312-19.

Beal, D.J. (1995). The Determination of Socially Optimal Recreational Outputs and Entry Prices for National Parks in Southwestern Queensland. PhD Thesis, The University of Queensland, Brisbane, Australia.

Bishop, R.C. and Heberlein, T.A. (1979). Measuring Values of Extramarket Goods: Are Indirect Measures Biased? *American Journal of Agricultural Economics*, 61: 926-930.

Blamey, R.K., Bennett, J.W., Louviere, J.J., Morrison, M.D., and Rolfe, J.C. (1997). The Use of Causal Heuristics in Environmental Choice Modelling Studies. Paper presented to the conference of the Australian and New Zealand Society for Ecological Economics, Melbourne, Australia, 17-20 November.

Bockstael, N. E., McConnell, K. E., and Strand, I. E. (1991). Measuring the Demand for Environmental Quality. *Contributions to Economic Analysis*, no. 198, pp. 227-70. North-Holland, Amsterdam.

Cameron, T. (1992). Combining Contingent Valuation and Travel Cost Data for the Valuation of Nonmarket Goods. *Land Economics*, 68:302-317.

Carson, R., Flores, N., Martin, K., and Wright, J. (1996). Contingent Valuation and Revealed Preference Methodologies: Comparing the Estimates for Quasi-Public Goods. *Land Economics*, 72:80-99.

Clawson, M. (1959). *Methods of Measuring the Demand for and Value of Outdoor Recreation*. Reprint No. 10, Resources for the Future, Washington, D.C.

Davis, R.K. (1963). The Value of Outdoor Recreation: An Economic Study of the Maine Woods. PhD Thesis, Harvard University, Cambridge.

Drimil, S.M. (1996). Sustainable Tourism in Protected Areas? An Ecological Economics Case Study of the Wet Tropics World Heritage Area. PhD Thesis, Australian National University, Canberra.

Fisher, A.C. and Hanneman, W.M. (1987). Quasi-Option Value: Some Misconceptions Dispelled. *Journal of Environmental Economics and Management*, 14: 183-90.

Hufschmidt, M.M., James, D.E., Meister, A.A., Bower, B., and Dixon, J.A. (1993). *Environment, Natural Systems and Development: An Economic Valuation Guide*. The Johns Hopkins University Press, Baltimore.

Hundloe, T., Mcdonald, G. and Blamey, R. (1990). *Socioeconomic Analysis of Non-Extractive Natural Resource Use in the Great Sandy Region*. A report to the Queensland Department of Environment and Heritage, Institute of Applied Environmental Research, Griffith University, August.

Kask, S.B. and Shogren, J.F. (1994). Benefit Transfer Protocol for Long-Term Health Risk Valuation: A Case of Surface Water Contamination. *Water Resources Research*, 30: 2813-2823.

Kahneman, D. and Knetsch, J.L. (1992). Valuing Public Goods: The Purchase of Moral Satisfaction. *Journal of Environmental Economics and Management*, 22: 57-70.

Kling, C.L. (1997). The Gains from Combining Travel Cost and Contingent Valuation Data to Value Non-market Goods. *Land Economics*, 73(3):428-439.

Lancaster, K.J. (1966). A New Approach to Consumer Theory. *Journal of Political Economy*, 74: 132-57.

Loomis, J.B. (1992). The Evaluation of a More Rigorous Approach to Benefit Transfer: Benefit Function Transfer. *Water Resources Research*, 28: 701-705.

Louviere, J.J. and Woodworth, G. (1983). Design and Analysis of Simulated Consumer Choice or Allocation Experiments: An Approach Based on Aggregate Data. *Journal of Marketing Research*, 20: 350-367.

Mackenzie, J. (1993). A Comparison of Contingent Preference Models. *American Journal of Agricultural Economics*, 75: 593-603.

Morrison, M.D., Bennet, J.W., and Blamey, R.K. (1998). *Valuing Improved Wetland Quality Using Choice Modelling*. Choice Modelling Research Report No. 6.

School of Economics and Management, University College, The University of New South Wales.

National Oceanic and Atmospheric Administration, NOAA (1993). Appendix I-Report of the NOAA Panel on Contingent Valuation. *Federal Register*, 58(10): 4602-4614.

Nelson, J.P. (1982). Highway Noise and Property Values: A Survey of Recent Evidence. *Journal of Transport Economics and Policy*, XVI: 117-138.

Parsons, G.R. and Kealy, M.J. (1994). Benefits Transfer in a Random Utility Model of Recreation. *Water Resources Research*, 30(8): 2477-2484.

Ready, R.C., Whitehead, J.C. and Blomquist, G.C. (1995). Contingent Valuation when Respondents are Ambivalent. *Journal of Environmental Economics and Management*. 29: 181-196.

Reiling, S.D., Boyle, K.J., Philips, M.L. and Anderson, M.W. (1990). Temporal Reliability of Contingent Values. *Land Economics*, 66(2): 128-134.

Soccimarro, M. (1992). An Analysis of User Pays for Queensland National Parks. B.ECON Honours Thesis, Department of Economics, The University of Queensland, Brisbane, Australia.

Stoeckl, N. (1994). A Travel Cost Analysis of Hinchinbrook Island National Park. Paper presented to the Tourism Research National Conference, 10-11 February, Gold Coast.

Smith,V.K. and Kaoru, Y. (1990). What Have We Learned Since Hotelling's Letter?: A Meta-Analysis. *Economic Letters*, 32: 267-72.

Teisl, M.F., Boyle, K.J., McCollum, D.W. and Reiling, S.D. (1995). Test-Retest Reliability of Contingent Valuation with Independent Sample Pretest and Post-Test Control Groups. *American Journal of Agricultural Economics*, 77: 613-619.

Ward, F.A. and Beal, D. (2000). *Valuing Nature with Travel Cost Models*. Edward Elgar, Cheltenham, U.K.

Ward, F.A. and Loomis, J.B. (1986). The Travel Cost Demand Model as an Environmental Policy Assessment tool: A Review of the Literature. *Western Journal of Agricultural Economics*, 11(2):164-78.

6.   Cost-Benefit Analysis

Objectives

After studying this chapter you should be in a position to:

❑   explain the conceptual basis of cost-benefit analysis derived from the market model;

❑   identify and value various types of costs and benefits;

❑   calculate discounted cash flows for costs and benefits;

❑   calculate project performance criteria; and

❑   explain how sensitivity analysis and risk analysis are conducted.

6.1   Introduction

In Chapter 3 we discussed how markets are supposed to work and in Chapter 4, we discussed why markets fail. Market failure implies either that most environmental goods are unpriced or underpriced. In Chapter 5, we considered methods that can enable us to place dollar values on environmental goods and services. In this chapter, we bring together concepts from the previous three chapters to form a framework for policy decision-making. This approach is formally referred to as cost-benefit analysis (Box 6.1). The chapter begins with a brief discussion of the conceptual basis of cost-benefit analysis. This is followed by a description of the steps in a CBA. The steps are illustrated with a real-life case study.

Box 6.1 A short history of cost-benefit analysis

The first formal application of CBA was in 1768 to evaluate the net benefits of the Forth-Clyde canal in Scotland. CBA first received official government recognition under the U.S. Flood Control Act of 1936. Under this act the United States Army Corps of Engineers were required to evaluate the benefits and costs of all water resource projects to whoever they accrue. The aim was to show that flood control was in the interests of social welfare. In 1950, the U.S. Federal Inter-Agency River Basin Committee introduced the *Green Book* which attempted to set out a procedure for comparing costs and benefits. In 1958, economists such as Jack Krutilla and Eckstein introduced the concepts of opportunity costs and Pareto optimality into the evaluation of water resource developments. In the U.K., formal recognition of CBA was given in 1967 (Pearce, 1983).

In the sixties and seventies, attention was given to the use of CBA in evaluating projects in developing countries. Different procedures were proposed by the Organisation for Economic Cooperation and Development (Little and Mirlees, 1974) and the United Nations Industrial Development Organisation (Dasgupta *et al.*, 1972). Subsequently, the World Bank suggested a unified approach (Squire and Tak, 1975).

Today, CBA is used to evaluate choices between alternatives in a wide range of areas including health, transportation, natural resources and agriculture, business, mining and mineral exploration, and many others.

Sources: Dasgupta *et al.* (1972); Little and Mirlees (1974); Squire and Tak (1975); Pearce (1983).

6.2   Utility, Benefits and Costs

The basis of 'value' in economics is the satisfaction or 'utility' a consumer derives from consuming a given good or service. On the basis of the utility which a consumer expects to gain from consuming the good, a person may be willing to pay for that good.

6.2.1 Willingness-to-Pay and Consumes Surplus as Measures of Benefits

In Chapter 3, the concept of the demand curve was introduced. According to the law of demand, more of a good or service will be purchased as its price falls. The converse is that less will be purchased as the price rises. This law holds given fixed incomes, tastes and preferences. The demand curve is also referred to as the **marginal benefit curve** because it indicates the benefit of consuming one extra unit of a good. The marginal benefit curve therefore provides us with an idea of changes in 'utility' or level of satisfaction. The negative slope of the demand curve reflects **diminishing marginal utility**. That is, as more units of a good are consumed, the extra satisfaction obtained from consuming an additional unit declines.[36] The **marginal rate of substitution** between two goods q_1 and q_2, MRS_{q1q2}, is equal to the ratio of the two marginal utilities (MU). Utility is maximised when the ratio of the marginal utilities are equal to the ratio of the prices for the two goods. That is,

$$MRS_{q_1q_2} = \frac{MU_1}{MU_2} = \frac{p_1}{p_2} \tag{6.1}$$

Cross-multiplying and rearranging, we can write:

$$p_1 = MU_1 \frac{p_2}{MU_2} \tag{6.2}$$

This equation can be interpreted as follows: assuming the price (p_2) and the marginal utility (MU_2) of other goods remain constant, the price of a good (p_1) is proportional to the satisfaction or utility of an additional unit (MU_1) of the good.[37] All this, of course, is an abstraction and depends on a perfectly competitive market. Nevertheless, the important point here is that the price one is willing to pay for a good depends on the satisfaction one derives from consuming it. This is taken to be a measure of benefits. However, for some

---

[36] It is important to note that although total utility increases, marginal utility declines (i.e., total utility increases at a declining rate).

[37] One important assumption made here is that the utility gained from each dollar spent (the marginal utility of income) is constant.

goods (especially environmental goods), the benefit or willingness-to-pay exceeds the market price, if one exists. In such cases, therefore, the correct measure of total benefits is the total revenue (price multiplied by number of units sold), area $0p_1bq_1$, plus the consumer surplus, triangle ap_1b (Figure 6.1).

The valuation methods that were discussed in Chapter 5 are concerned with obtaining estimates of WTP. Although the WTP concept is useful in applied work, it is instructive to draw attention to two problems associated with it.

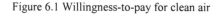

Figure 6.1 Willingness-to-pay for clean air

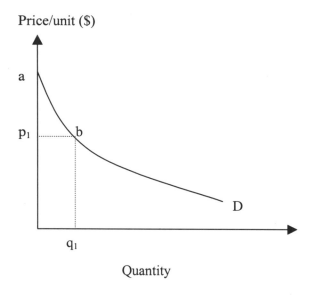

1. WTP does not fully reflect **intensity of preference**. Although WTP depends on income, a person who is willing to outlay all of his or her income on a good may have a higher intensity of preference compared to another who is willing to pay a similar price but the amount is a small fraction of his or her income.

2. The WTP concept assumes that all persons in the population have the same **marginal utility of income**.[38] However, this is not realistic. One dollar may yield more utility to a poor man than to a rich man. However, the WTP concept assigns the same weight to the $1 in both instances.

6.2.2 The Concept of 'Cost' in CBA

The objective for measuring costs is to obtain information that will enable us to carry out a cost-benefit analysis of alternative projects. Given that resources are scarce, the selection of a given investment means that the resources will not be available for an alternative use. Thus, there is an opportunity cost (OC) to carrying out the investment. Opportunity cost is defined as the benefits that would have been obtained from the forgone alternative investment. If there are well functioning markets (i.e., perfect competition), then the opportunity cost of a good is the market price for that good. However, in practice, markets are distorted for some goods or, in the case of environmental goods, are non-existent. In such cases, we have to use alternative means of estimating the opportunity costs. These methods are discussed later.

6.2.3 The Concept of Net Social Benefits

The objective of social CBA is to determine whether a project is socially beneficial, i.e., whether the net social benefits (NSB) are positive. Using the concepts introduced above,

$$NSB = WTP - OC \qquad (6.3)$$

The rationale of CBA is that if NSB is positive then, in theory, the state can use the surplus to compensate the losers. That is, a project is socially desirable if it can result in a potential Pareto improvement. This rule has generated some controversy because it does not require that actual compensation take place. Although, increasingly, we are seeing cases in which government is required, sometimes through the court system, to actually compensate losers in big resource development projects.

---

[38] Marginal utility of income is the change in utility for a unit change in income.

We need to make a distinction between social CBA, which is carried out from society's perspective, and private CBA, which is carried out from an individual investor's viewpoint. The former is formally referred to as an **economic analysis**, whereas the latter is referred to as **financial analysis**. For example, a project could be financially viable, from an investor's point of view, but be socially undesirable due to, say, adverse environmental impacts. The process of conducting a social CBA involves the following steps.

- Define the objectives and the scope of the project
- Identify and screen the alternatives
- Identify and value the costs and benefits for the remaining alternatives
- Calculate discounted cash flows and project performance criteria for each alternative
- Rank the alternatives in order of preference
- Conduct a sensitivity analysis and/or risk analysis for the preferred alternative(s), and
- Make a final recommendation

These steps are briefly discussed in the following sections. To illustrate the process of conducting a CBA, we shall make use of the Bintuli Wastewater Treatment case study (Box 6.2).[39]

6.3   Defining Objectives and Project Scope

Every project must have an objective (or objectives). The objective is often specified by decision-makers in the bureaucracy. However, to facilitate the process of the CBA this objective (or objectives) should be clear and unambiguous. In our illustrative case study, The Bintuli Wastewater Treatment Project (Box 6.2), the project's objectives are to improve the health of the community and increase economic activity by improving wastewater treatment facilities in the city.

---

[39] Fictitious names have been used in this case study to protect confidentiality.

Box 6.2 Case Study 1: The Bintuli Wastewater Treatment Project

The city of Bintuli is a thriving centre of commerce and industry in the Republic of Kabastan. The main industries in the area include metal manufacturing, coal extraction, chemical manufacturing, construction, papermaking and food processing. The value of industrial output in 1990 was estimated at $200 million while agricultural output was valued at $16 million. In recent years, the quantities of industrial and domestic effluent discharged into the Nombeng and Weban Rivers, as well as other streams and small rivers in the area, have increased rapidly. According to the Environmental Protection Agency (EPA), total industrial effluent in Bintuli is 163,700 m^3/day and total wastewater discharge, including domestic waste is 271,700 m^3/day. Thirty percent of the industrial effluent is treated by an existing on-site treatment system. There is no existing central wastewater treatment system for domestic waste. All domestic waste, as well as the untreated industrial effluent, is currently discharged into the nearby rivers.

It is proposed to build a wastewater treatment facility and associated pumping stations and drainage pipework. The facility will treat approximately 28 percent of industrial waste with the balance being domestic waste. The treated water of clean water standard will be discharged into the Nombeng River for use by industry and agriculture. Bintuli, like many cities in Kabastan, is facing serious depletion of groundwater resources and severe pollution of surface water due to insufficient infrastructure available to accommodate the rapidly increasing amounts of industrial and domestic effluent. Currently, approximately 70 percent of the city's effluent is dumped untreated into the Nombeng and other rivers in the area. The river courses in the city have turned black and emit unpleasant odours, creating health hazards.

Continued discharge of untreated wastewater at the current rates will, in time, render the surrounding areas non-viable in economic, as well as social terms. Prospects for sustainable economic development in the area critically depend on the level of wastewater infrastructure development.

6.4   Identifying and Screening the Alternatives

At this stage of the process, all possible options for achieving the objectives must be listed. One of the options must be the *status quo*, which is the 'do nothing' option. It must be borne in mind that the 'do nothing' option is not without costs. In this particular example, doing nothing about waste water treatment will result in social costs. Therefore, the avoidance of these costs must be counted among the benefits of the project. The other options include expanding the existing wastewater treatment facilities, and building a new wastewater treatment facility.

After screening the alternatives, expansion of the existing facility is ruled out because it uses outdated technology and would be expensive to maintain. Other options could include various locations or site options. For ease of presentation, only one potential site is considered in the discussion.

6.5   Identifying the Benefits and Costs

A 'benefit' in social CBA is an outcome which results in an increase in an individual's utility, whereas a 'cost' is an outcome which results in a decrease in an individual's utility. Once again, it is necessary to reiterate that this definition of costs and benefits will be different for a private investor whose objective is to maximise profits. In this case, a cost is any outcome that decreases profits and a benefit is any outcome that increases profits. Other important points to note include the following:

1. In assessing the project's contribution to the objective(s), we only consider additional (i.e., marginal) changes in costs and benefits and not the total costs and benefits. That is, we net out the costs and benefits **without** the project from the costs and benefits **with** the project. This is referred to as the **incremental approach**.

2. Point (1) implies that we must exclude sunk costs and benefits. That is, costs and benefits that are incurred before the commencement of the project. The rationale is that previous

costs, for example, are not an opportunity cost because they do not represent a loss of future income from an alternative use of the resource.

3. Transfer payments should be excluded. A transfer payment is money flow from one group in the community (e.g., the government) to another. Such payments should be excluded because they do not result in an increase in net benefits. Examples are taxes, subsidies, loans, and debt service (i.e., payment of interest and repayment of principal). Note, however, that taxes paid by foreign investors must be included because they increase net benefits.

Other cost items which are normally included in financial statements but excluded from social CBA are **depreciation** and **interest**.

4. Depreciation: As a rule, depreciation of capital cost items is not included as a cost item in social CBA. The process of discounting values the capital items at their opportunity costs over the life of the project. Therefore imputing depreciation as a cost would result in double counting.

5. Interest: The discount rate used in CBA already takes into account relevant factors including the interest rate. The discounting procedure reduces the stream of costs to their present values. Thus, once again, the inclusion of interest as a cost item would result in double counting.

Costs and benefits are normally classified into two groups: **primary** costs and benefits; and **secondary** costs and benefits. Primary costs and benefits arise directly from the project, while secondary costs and benefits arise from activities or events that are triggered by the project. For example, suppose a large-scale agricultural project would result in an increase in farm produce. In this case, the secondary benefits would include the increase in the profits of businesses that process agricultural products.

Secondary costs and benefits should be handled cautiously because they could exaggerate the estimates. As much as possible the opportunity cost principle must be used as a guideline. At issue is whether resources have been merely transferred from one part of the economy to another. A key question to ask is whether the resources employed would have had alternative uses. For example, if a project results in an increase in the number of employed people in a village, this would only be counted as a secondary benefit if there were no alternative employment opportunities. In this case, the opportunity cost of labour is zero.

Based on the discussion in the previous chapter, we can also divide costs and benefits into market and non-market costs and benefits. As already stated, non-market costs and benefits apply to goods and services which are not traded and for which market prices do not exist. As far as possible, non-market costs and benefits must be identified and valued.

6.5.1   *Identifying Costs and Benefits in Case Study 1: The Bintuli Wastewater Treatment Project*

The primary costs of the proposed project comprise the investment cost and operating and maintenance (O&M) costs. The investment costs include construction of a pumping station, office building and wastewater treatment facilities and purchase of equipment. The O&M costs include wages and salaries, fuel and chemical costs and other costs (e.g., project management, project preparation, training and commissioning).

Without an additional increase in wastewater treatment capacity, water pollution in Bintuli town will continue to increase and, in the long-run, impose severe economic and social costs on the community. Therefore, the net economic benefits from the project could be attributed to the avoidance of these costs. The economic benefits include the financial benefits derived from user charges and the economic benefits derived from wastewater treatment. The primary economic benefits are:

- a reduction in health costs and mortality rates due to reduced pollution to water resources and domestic drinking water;
- a reduction in the costs of treating increasingly polluted water supplies, and
- an increase in labour productivity as a result of a reduction in absence from work due to illness;

The secondary economic benefits from the project are:

- benefits to industry and agriculture from using recycled water;
- additional revenues from re-afforestation; and
- increase in reed harvesting for the paper mill industry.

6.6 Valuing the Costs and Benefits

Once costs and benefits have been identified, they must be valued in order to allow comparison between alternatives. The basic assumption here is that prices reflect value or opportunity costs, or can be adjusted to do so. As already suggested, the procedure is to value the costs and benefits according to the opportunity cost principle. This is in recognition of the fact that the market price of a good may not necessarily reflect its opportunity cost or scarcity value (i.e., the value of that good in its next best use). As was explained in Chapter 4, this difference between prices and value is a direct result of market failure. Thus, in a social CBA the prices of inputs (and outputs) which do not reflect their true value to the society are adjusted. This process is referred to as **shadow pricing**, and involves adjusting the market prices by given discount factors.

It is important to stress again that the correct procedure is to identify and value the costs and benefits that arise **with** the project and to compare them with the situation that would prevail **without** the project. The difference is the **net incremental benefit** arising from the project. The 'with' and 'without' comparison should not be confused with a 'before' and 'after' comparison. The comparison fails to account for changes in output that would occur without the project and thus could lead to an erroneous statement of the benefits derived from the project.

Consider the example of an investment project that increases exports by 3 percent. Assume that without the investment, exports will increase by 1 percent. Using the before and after approach, one could wrongly attribute the total increase in exports (i.e., 3 percent) to the project. Whereas, in actual fact, what could be attributed to the project investment is only the 2 percent incremental increase.

6.6.1 Valuing the Costs

The first step in valuing costs and benefits is to find market prices for the inputs and outputs. All costs must be in **present day** or **constant prices**.

That is, costs incurred over the project life must be valued at prices prevailing at the time of the project's appraisal. This approach assumes that annual costs will increase at the inflation rate, which implies that the cash flows are expressed in **real** rather than **nominal** prices.

Residual (or salvage) values

Some assets may have economic lives that exceed the planning horizon or project life, or may have reached their economic lives but still have a scrap value. The **economic life** is the estimate of the length of time for which it is economically viable to operate the asset without a major refurbishment. In such cases, the residual or salvage value of the asset must be included as a cash inflow (benefit) at the end of the planning horizon.

Calculation of residual value

The residual value (or salvage value) of an asset is normally assessed at a level pro rata to the remaining economic life. There are two major approaches to assessing residual value: the linear method and the diminishing value method.

The linear method

As the name suggests, this method assumes that the residual value declines linearly over time. The residual value at time t is given by:

$$(1 - td)P \tag{6.4}$$

where d = annual proportional decline in value = $1/n$, where n is the economic life; P is the initial price and t = time. For example, assume an asset is purchased at \$100,000 and has an economic life of 20 years. At the end of a planning period of 15 years, its residual value is given by:

$$(1 - 15 \times 1/20)100,000 = 0.25 \times 100,000 = \$25,000 \tag{6.5}$$

The diminishing value method

This approach assumes that the value of the asset declines by a fixed proportion of the beginning-of-year value per annum. The residual value at time t is given by:

$$(1 - d)^t P \qquad (6.6)$$

The value of d is normally taken to be about 1.5 times that used in the linear model. Using the above example, $d = 1.5 \times 0.05 = 0.075$. Thus, the residual value at time $t = 15$ years would be given by:

$$(1 - 0.075)^{15}\, 100{,}000 = \$31{,}055 \qquad (6.7)$$

Land and pre-existing buildings and plant

Land, buildings and plant already owned by the operating authority must be valued at their opportunity costs. These opportunity costs should be current valuations based on the most profitable alternative uses.

Staged construction

Where a project is to be implemented in stages, only the proportion of the investment and operating costs required to satisfy demand in the current planning horizon must be attributed to the project.

Working capital

Working capital is often required to meet financial transactions in the initial period of the project. The amount committed is often of the order of 50 percent of operating and maintenance costs or 2 percent of the total capital outlays. Working capital must be treated as a cash outflow at the time when capital expenditures are made, with the full amount being released as a capital inflow at the end of the project.

Operating costs

Operating costs typically occur every year and include the following: labour; utilities; supplies; repairs and maintenance; equipment hiring and leasing; insurance and administrative overheads. These items are to be estimated on an annual basis.

Implicit costs

In addition, there could be implicit or opportunity costs and social costs associated with a project. The opportunity costs arise with respect to the use of land, buildings, plant, and machinery already purchased by the local authority. Implicit costs could also arise with respect to time spent on the project by the authority's staff and management.

The above items may not involve the use of cash outlays but, nevertheless, represent a cost because the resources they tie up could have been used for other purposes. For example, in the case of the Bintuli Wastewater Treatment Project, the land used to store the sludge is likely to have an alternative use and therefore has an opportunity cost.

Valuing costs in Case Study 1: The Bintuli Wastewater Treatment Project

All the equipment and construction materials are imported and are valued in US dollars. Fuel and chemical supply are adjusted by subtracting the government subsidies on these items. Because there is a high level of unemployment in the area, unskilled labour is shadow-priced at 50 percent of the going wage rate. Skilled labour is valued based on annual salaries. The total investment cost is estimated at $16.57 million and the operating and maintenance costs are $1.62 million per annum (Table 6.1). The construction of the project is expected to take three years.

6.6.2   Valuing the Benefits

The benefits, in the case of Bintuli Wastewater Project, include the revenues raised from user charges and the economic benefits derived from treating wastewater. The economic benefits include the following: reduced mortality; productivity gains from reduced morbidity; water treatment cost savings; sale of recycled water; afforestation benefits and reed harvesting. Each of these benefits are discussed below.

User charges

The determination of user charges was based on the principle of full cost recovery. That is, it was based on a level of charges that would recover the investment and O&M costs over the life of the project. This level was estimated to be 6.9 cents/m^3, of which operating and maintenance costs account for 2.96 cents/m^3. With the project, 54.75 million m^3/year of effluent would be treated, resulting in revenue of $3.78 million per annum.

Table 6.1 Investment and operating and maintenance costs, Bintuli
Wastewater Treatment Project

Item	Cost ($ million)
Investment Costs:	
Buildings and structures	3.42
Equipment and supplies	13.15
Total investment cost	16.57
Operating and Maintenance Costs:	
Electricity	0.68
Salaries	0.09
Chemicals	0.06
Maintenance	0.58
Other	0.21
Total O&M costs	1.62

As indicated above, 11.4 million m^3/year of industrial effluent is already being treated by the town council. The appropriate user charges for this wastewater treatment was determined to be 5.31 cents/m^3, providing revenue of $605,340 per annum. Based on this information, net incremental sales revenue (i.e., 'with project' sales minus 'without project' sales) would be $3.17 per annum by Year 6 when the new plant is at full capacity.

Estimation of the economic benefits
The estimates of the economic benefits are detailed in Table 6.2.

1. Recycled water benefits
About 60 percent of the town's wastewater will be recycled and used for irrigation and industrial purposes. The opportunity cost of this recycled water was estimated to be 10 cents/m^3. The economic benefits from recycled water were therefore estimated to be $66,000 in Year 4, rising to $3.29 million per annum by Year 8 (Table 6.2).

2. Afforestation benefits
Pine and some hard wood species will be planted on 142.8 ha of land. The net returns per hectare were taken from estimates provided by the Kabastani authorities for experimental plots. These were reported to be $689/ha. Applying

this figure to the proposed area resulted in net benefits of $10,000 in Year 8 and reaching a maximum of $100,000 by Year 17.

3. Reed harvesting
Economic benefits are also expected from reed harvesting for the paper mill industry. The Kabastani authorities have estimated the net returns to be $258.40 per hectare. Applying this figure to a projected area of 95.25 ha resulted in net annual benefits of about $20,000 from Year 6.

4. Reduced mortality benefits
Using World Bank estimates for Kabastan, mortality reduction from the project was taken to be 0.005, 0.008 and 0.024 percent, respectively, for the age categories 15–24 years, 25–59 years and over 60 years. On the basis of estimates for the number of people in each of these age categories, the total number of deaths saved per annum was calculated. Using the estimated proportion of people employed in each age category and the mortality reduction rates, an estimate of both employed and unemployed deaths was made.

Given the local annual wage of $620 (which includes housing subsidies, and other government payments) and assuming average working lives of between 5 and 45 years for the three age categories, annual income losses avoided were estimated. For the unemployed, a leisure value of half the annual wage was assumed.

On the basis of this estimate the annual gains in leisure from saving deaths were estimated. Given that the project will treat about half of Bintuli's wastewater, only 50 percent of the potential mortality and morbidity benefits were attributed to the project. The income benefits from reduced mortality were therefore estimated to be $10,000 in Year 4, rising to $110,000 by the end of the project (Table 6.2).

5. Productivity gains from reduced morbidity
A major social impact of the project is the reduction of the incidence of pollution-related illness and hence a reduction in worker absenteeism. These benefits were estimated as follows.

Table 6.2 Incremental economic benefits, Bintuli Wastewater
Treatment Project ($ millions)

Year (A)	Recycled water (B)	Afforestation (C)	Reed harvesting (D)	Reduced mortality (E)	Reduced morbidity (F)	Water treatment cost savings (G)	Incremental economic benefits $H = \sum (B \text{ to } G)$
1							
2							
3							
4	0.66			0.01	0.18	0.11	0.96
5	0.99			0.02	0.38	0.11	1.50
6	1.64		0.02	0.03	0.57	0.11	2.38
7	2.63		0.02	0.05	0.76	0.11	3.57
8	3.29	0.01	0.02	0.06	0.95	0.11	4.44
9	3.29	0.02	0.02	0.06	1.00	0.11	4.50
10	3.29	0.03	0.02	0.06	1.06	0.11	4.57
11	3.29	0.04	0.02	0.07	1.11	0.11	4.63
12	3.29	0.05	0.02	0.07	1.17	0.11	4.71
13	3.29	0.06	0.02	0.07	1.23	0.11	4.78
14	3.29	0.07	0.02	0.08	1.30	0.11	4.86
15	3.29	0.08	0.02	0.08	1.37	0.11	4.94
16	3.29	0.09	0.02	0.09	1.44	0.11	5.03
17	3.29	0.10	0.02	0.09	1.51	0.11	5.12
18	3.29	0.10	0.02	0.10	1.59	0.11	5.20
19	3.29	0.10	0.02	0.10	1.68	0.11	5.29
20	3.29	0.10	0.02	0.11	1.76	0.11	5.38

First, it was assumed that the current average number of days lost per worker per annum as a result of illness is 3 days. Next, using the employment statistics, potential productivity losses avoided per worker per annum were estimated to be about $180,000 in Year 4, rising to about $1.8 million by the end of the project.

6. Water treatment cost savings
As indicated above the benefits of the project include the avoided costs of treating polluted water. An estimate of $0.002/m^3 for water treatment cost

savings, consistent with World Bank estimates for Kabastan, was assumed. The economic benefits of water treatment cost savings were therefore estimated to be $110,000 per annum.

Summary of economic benefits

The incremental economic benefits derived from the project were estimated to be $96,000 in Year 4, rising to $5.38 million by the end of the project (Table 6.2).

6.7   Calculating Discounted Cash Flows and Project Performance Criteria

Once the costs and benefits with and without the project have been identified and valued (in monetary terms), the analyst is now ready to compare the costs and benefits in order to make a decision as to which alternative(s) to accept or reject.

Project performance criteria (or project selection criteria) provide a means by which different alternatives that last several years into the future and which have different streams of costs and benefits could be compared. Calculating these measures involves using the technique of **discounting**. We often use expressions like 'time is money', and 'a bird in hand is worth two in the bush', and so on. These expressions imply that money has a time value. In particular, present values are better preferred than future values. Calculating **discounted cash flows** (DCFs) involves 'reducing' future streams of benefits and costs to their 'present values' to enable comparisons to be made between competing alternatives.

For example, given an investment stream of:

t (years)	0	1	2
Investment	−$100	$50	$150

The net present value (NPV) at an interest rate of 10 percent is given by:

$$NPV = -100 + 50/(1 + 0.1)^1 + 150/(1.01)^2 = \$96.54 \qquad (6.8)$$

In more general terms, given a stream of benefits, B_0, B_1...B_n, and a stream of costs C_0, C_1,...C_n, the net present value (or net present worth) could be written as:

$$NPV = B_0 - C_0 + \frac{B_1 - C_1}{1+r} + \frac{B_2 - C_2}{(1+r)^2} + ... \frac{B_n - C_n}{(1+r)^n} = \sum_{t=0}^{n} \frac{B_n - C_n}{(1+r)^n} \quad (6.9)$$

6.7.1 Choice of Discount Rate

The discount rate in social cost-benefit analysis reflects society's preferences between present and future consumption. In general, a high discount rate implies that the society has a relatively stronger preference for present consumption over future consumption, while a low discount rate implies a relatively stronger preference for the present.

The choice of a discount rate in social CBA is a controversial one. Many environmentalists argue against discounting, in general, and high discount rates, in particular, because they believe high discount rates are associated with environmental degradation (Goodin, 1982). High discount rates are thought to be a cause of degradation because individuals prefer short-term measures to satisfy immediate needs at the expense of environmental conservation (Pearce, 1987; Pearce and Markandya, 1990). In the case of developing countries, it is often assumed that there is a vicious circle between poverty and environmental degradation. That is, high discount rates cause environmental degradation, which in turn worsens poverty and further increases discount rates. However, the adverse impacts of high discount rates on environmental quality have not been established beyond doubt. For example, it is possible that demand for natural resources could be lower at high discount rates (see Krautkraemer, 1985, for a proof). Furthermore, there have been instances where people facing imminent danger have taken decisions with long-term implications for their (or their family's) survival.

Returning to the practical issue of choosing a discount rate, economists have tended to use long-term interest rates on government bonds as one measure of the opportunity cost of capital. U.S. Government agencies use a discount rate of 10 percent. The Australian government recommends a rate of 8 percent for public projects (Department of Finance, 1991). The rate used must be the **real rate**. That is, the nominal interest rate minus the inflation rate. Usually, the discount rate comprises a **risk-free rate** and a **risk**

margin. For example, a real rate of 8 percent might include a risk-free rate of 5 percent and a risk margin of 3 percent.

A **sensitivity analysis** should be carried out to test the effect on the project selection criteria of using slightly higher and lower rates (e.g., 6 percent and 10 percent).

6.7.2   Period of Analysis

Another critical issue in computing DCFs is the **planning period** or **horizon**. This is the number of years for which cost and revenue data will be collected. The planning period will vary depending on the nature of the project. In general, the planning period should be determined by a period within which predictions could be made with a high degree of confidence. The planning period must, wherever possible, also correspond to the **economic life** of the project.

6.7.3   Choice of Project Performance Criteria

Project performance criteria include the following: **net present value** (NPV), **benefit-cost ratio** (BCR), **internal rate of return** (IRR), and **payback period**. NPV has already been defined above. The BCR is the ratio of the present value of project benefits to the present value of costs. The IRR is defined as the discount rate at which the present value of project benefits equals the present value of costs. The BCR is defined as:

$$BCR = \frac{B_0 + \dfrac{B_1}{1+r} + \dfrac{B_2}{(1+r)^2} + \ldots \dfrac{B_n}{(1+r)^n}}{C_0 + \dfrac{C_1}{1+r} + \dfrac{C_2}{(1+r)^2} + \ldots \dfrac{C_n}{(1+r)^n}} = \frac{\displaystyle\sum_{t=0}^{n} B_n / (1+r)^n}{\displaystyle\sum_{t=0}^{n} C_n / (1+r)^n} \qquad (6.10)$$

The IRR can be found by finding the discount rate (i) at which the following equations holds:

$$B_0 - C_0 + \frac{B_1 - C_1}{1+i} + \frac{B_2 - C_2}{(1+i)^2} + \ldots \frac{B_n - C_n}{(1+i)^n} = 0 \qquad (6.11)$$

The IRR may not exist or may not be unique.

The payback period is defined as the number of years required for a project to recover its costs. In general, the payback period discriminates against projects with high capital expenditures and long-term benefits. It is not recommended as a measure of project worth.

Although NPV and BCR can easily be calculated with the help of a calculator, IRR is more difficult to calculate. A trial and error method must be used to compute Equation (6.11) using different values of i. All three measures are easily obtained using spreadsheet programs such as Excel, Lotus and Quattro-Pro. An example is provided below from Case Study 1.

The decision rule is to accept a project when NPV $\geq$ 0, BCR $\geq$ 1 and the IRR exceeds the social opportunity cost of capital. Equations (6.9) and (6.10) indicate that when NPV = 0, then BCR = 1. When evaluating a single alternative, all three measures will yield the same result. However, when used to rank several alternatives, the three measures can yield different results.[40]

NPV is the most preferred performance criterion in social cost-benefit analysis because it provides an estimate of the size of the potential Pareto improvement, the basis of the CBA approach. If two or more projects have positive NPVs, the IRRs can be used to rank them. The IRR describes the rate at which a project transforms present benefits into future benefits. It is the maximum interest rate at which a given project could recover the investment and operating costs and still break even.

Calculating DCFs and project performance criteria for Case Study 1, Bintuli Wastewater Treatment Project

In this section, we will use the information produced from Section 6.6 to calculate discounted cash flows for the Bintuli Wastewater Treatment Project. Issues that need to be resolved before the DCFs can be calculated are: (1) choice of discount rate and, (2) the planning period.

1. Choice of discount rate: a real rate of 12 percent will be used to produce the DCFs. This rate is the average of published World Bank discount rates for the last 10 years for this country.

---

[40] See Exercises 1 to 4 at the end of the chapter.

2. Planning period: a planning period of 20 years is used. Based on advice received from engineers, this is a reasonable period given the type of equipment to be used.

Incremental net economic costs

From Table 6.1, the total investment cost of the project is $16.57 million and is spread over the construction period of three years as follows: $2.01, $8.45 and $6.11 million, respectively. The cost of operating the current facility (the without project option) is $0.31 million per annum. The total O&M costs are therefore

> Box 6.3 Excel hints for carrying out the computations in Table 6.3
>
> Incremental net benefits:
> (i)   Year 1: place your cursor in cell *E5*. Type =*(C5+D5−B5)*.
> (ii) Copy the formula in cell *E5* by clicking on the 'copy' icon on the toolbar.
> (iii) Paste this formula into rows *E6* to *E24*.
>
> Calculating NPV:
> Place your cursor in an appropriate place, say, cell *E25*. Type:
> =*NPV(12%, E5:E24)*
> where 12% is the discount rate.
>
> Calculating IRR:
> Place your cursor in an appropriate place, say, cell *E26*. Type:
> =*IRR(E5:E24)*

$1.93 million per annum and the net incremental O&M costs (i.e., 'with' minus 'without' projects costs) are therefore $1.62 million per annum. The net incremental economic cost of $2.42 million in Year 4 includes allowance for a working capital of $0.81 million, equivalent to 50 percent of O&M (Table 6.3).

Incremental net benefits

The incremental sales revenue has been estimated to be $1.91 million in Year 4, rising to $3.17 million in Year 6. The economic benefits are $960,000, commencing in Year 4. The incremental net benefits, Columns C plus D minus Column B, are therefore negative $2.0 million in Year 1 and rise to $9.8 million by the end of the project (see Excel hints in Box 6.3). The NPV at the 12 percent discount rate is $12.08 million (Table 6.3).

The IRR is 21 percent, which is above the opportunity cost of capital of 12 percent. Therefore, it can be concluded that the Bintuli Wastewater Treatment Project is economically viable.

Table 6.3 Net incremental economic benefits, Bintuli Wastewater
Treatment Project ($ million)

Year (A)	Incremental economic costs (B)	Incremental sales Revenue (C)	Incremental economic benefits (D)	Incremental net benefits E = (C + D) − (B)
1	2.01			−2.01
2	8.45			−8.45
3	6.11			−6.11
4	2.42	1.91	0.96	0.45
5	1.62	1.91	1.50	1.80
6	1.62	3.17	2.38	3.93
7	1.62	3.17	3.57	5.12
8	1.62	3.17	4.44	5.99
9	1.62	3.17	4.50	6.05
10	1.62	3.17	4.57	6.12
11	1.62	3.17	4.63	6.19
12	1.62	3.17	4.71	6.26
13	1.62	3.17	4.78	6.34
14	1.62	3.17	4.86	6.41
15	1.62	3.17	4.94	6.50
16	1.62	4.43	5.03	7.84
17	1.62	4.43	5.12	7.93
18	1.62	4.43	5.20	8.01
19	1.62	4.43	5.29	8.10
20	-0.02	4.43	5.38	9.83
			NPV @ 12%	$12.08
			IRR	21%

6.8   Concepts of Risk and Uncertainty

In economics and business, a distinction is made between **risk** and **uncertainty**.[41] The term 'risk' is used to refer to a potential outcome (or outcomes) where the magnitude of the outcome (or outcomes) is known and the probability of occurrence is known or can be determined. On the other hand, 'uncertainty' is used to refer to a situation where the magnitude of the outcome may or may not be known and the probability of occurrence is unknown. In practical situations, however, it is difficult to define precisely the probability of an occurrence. Therefore, the distinction between risk and uncertainty may not be clear-cut.

In some cases, the degree of risk can be assessed based on past records. For example, the probability of a cyclone occurring is generally available from climatic data, and the probability of a flood may be obtained from hydrological data. In such cases, one may be able to predict the probability of a cyclone or flood. However, such forecasts can be imprecise because major floods, for example, do not occur frequently and therefore the probability estimates can be unreliable.

There are a number of reasons why there would be uncertainty about any large-scale public project. Such projects tend to be long-lived, occurring over, say, 20 to 30 years. The valuation of some items of costs and benefits (especially environmental impacts) will always be uncertain due to lack of reliable data. Finally, it is difficult to forecast the nature of the operating environment 20 to 30 years into the future. Nevertheless, because CBA requires predictions about risk to be made, there is need for a structured approach to assess the extent of the risk.

The aim of risk analysis is to enable measures to be taken to reduce the risk at the project design phase. Possible measures could include spreading orders among several suppliers, using alternative fuels, and trading off lower efficiency for higher reliability. The common methods of accounting for risk and uncertainty in CBA include sensitivity analysis, break-even analysis, switching or cross-over values, and risk analysis. These methods are briefly discussed below.

---

[41] For a more detailed discussion of risk and uncertainty, readers may refer to the following sources: Hertz and Thomas (1983), Megill (1985) and Morgan and Henrion (1990).

6.9   Sensitivity Analysis

Sensitivity analysis is used to assess the possible impact of uncertainty by posing 'what if' questions. These questions pertain to what would happen to the project's viability if some or all of the key parameter values happen to be different from the original values. A sensitivity analysis highlights the critical factors affecting the project's viability. This allows the decision-makers or project manager to pay attention to these factors during the implementation stage. Parameters subjected to sensitivity analysis include:

- the discount rate
- length of the project planning horizon
- different timing of the project's operation
- changes in the capital outlays
- changes in the price of non-market goods, and
- changes in social and environmental benefits and costs

The sensitivity analysis is normally carried out by recalculating the project performance criteria, using a range of values for the uncertain parameter (or parameters). The results of the sensitivity analysis may be in the form of two-way and three-way tables. A sensitivity analysis helps the project analyst to identify the range of parameter values within which a project can remain economically viable. It helps to identify critical variables and, in so doing, provides information that could be used to redesign a particular option. Sensitivity analysis may also point to the need to acquire additional information to ensure that the assumptions are more realistic.

The project performance criteria commonly used in sensitivity analysis are NPV and IRR. The procedures in a sensitivity analysis are as follows:

1.   Determine a realistic range of values for the variables that are subject to uncertainty. For example,

- Capital cost, ± 30 percent
- O&M costs, ± 30 percent
- Product prices, ± 30 percent

2.   Calculate the effect of possible changes on the project selection criteria, while varying one variable and holding the others constant.

3. Re-consider the economic viability of the project (i.e., the 'robustness' of the project profitability) in light of the calculations in (2) above.

Although a number of variables that can be subjected to sensitivity analysis have been listed above, there is no point in examining all of them arbitrarily. Given the usual constraints in conducting such analyses, only the key critical variables should be analysed.

6.9.1   Break-even analysis

For a given project or option, the **break-even value** is the value of the discount rate at which the NPV is zero. This is the value at which the entire project's costs can be recovered.[42] For some projects where the main benefits are environmental protection, it may not be possible to use break-even analysis. On the benefit side, if a variable (e.g., price) appears to be higher than the break-even level, that increases confidence in the project's viability. Similarly, on the cost side, an estimate which is lower than the break-even level would suggest that the project is likely to be economically viable.

6.9.2   Switching (cross-over) values

The **switching** or **cross-over** value of a project performance criterion (e.g., NPV) is the discount rate at which the ranking of two projects changes. Switching or cross-over values are recommended when considering only one uncertain variable. The intention is to determine the value at which the NPV becomes zero, or the value at which two alternatives change rank. The next step is to indicate the chances of the NPV taking on values above or below the switching value in order to gain an idea of the riskiness of the project.

6.9.3   Conducting sensitivity analysis for Case Study 1, Bintuli Wastewater Treatment Project

For the Bintuli Wastewater Treatment Project, the critical uncertain variables chosen for analysis were:

---

[42] A good reference on break-even analysis is Schweitzer *et al.* (1986).

- changes in the capital and O&M costs, and
- changes in the net incremental economic benefits

These variables were subjected to changes within the ±30 percent range. The results indicate that the IRR is robust (Table 6.4). A 30 percent decline in economic benefits reduces the IRR to 17 percent, assuming no change in capital costs and no change in O&M costs. A 30 percent increase in capital costs, assuming no change in economic benefits, reduces the IRR to 17 percent. Similarly, a 30 percent increase in operating costs reduces the IRR to 19 percent. These results indicate that the estimate is insensitive to large changes in the projected economic benefits and costs.

Table 6.4 Sensitivity of the internal rate of return, Bintuli Wastewater Treatment Project

		Change in Net Economic Benefits				
		−30%	−15%	0%	+15%	+30%
	−30%	23	25	27	29	31
Change in	−15%	20	22	24	25	27
Capital	0%	17	19	21	23	24
Costs	+15%	15	17	19	21	22
	+30%	14	16	17	19	20
	−30%	19	21	23	25	26
Change in	−15%	18	20	22	24	25
O&M	0%	17	19	21	23	24
Costs	+15%	16	18	20	22	23
	+30%	15	17	19	21	22

6.10   Risk Analysis

As indicated above, sensitivity analysis is suitable when only a few parameters are uncertain. In cases where the values of several parameters are uncertain, risk analysis based on probabilities of the key variables should be undertaken. In this approach, the probabilities of occurrence of the key variables are used as weights to recompute the project performance criteria.

Risk analysis can be carried out using special purpose computer packages such as @RISK (Palisade Corporation, 1997) which is available as an add-in to spreadsheet packages (e.g., Lotus 1-2-3 and Microsoft Excel). Through repeated iterations, the @RISK program generates probability distributions for NPV and IRR. The result is a cumulative probability graph which indicates, for example, the probability of the NPV falling below zero. For both NPV and IRR, useful statistics such as the mean, maximum, minimum, standard deviation and coefficient of variation are also provided.

The major practical difficulty in conducting risk analysis is in obtaining probability estimates. In some cases it may be possible to obtain information about the chances of an event occurring. For example, it is possible to obtain information about the frequencies of road accidents, floods, outbreaks of diseases, and so on. Probabilities can be calculated from such historical data. Where there is little or no historical data, a subjective assessment will have to be made.

Although a wide range of theoretical probability distributions is available, it is preferable to use the most 'accurate' probability distribution. Very often it is not possible to obtain a predefined probability distribution due to lack of data and therefore an assumed distribution is used. Common distributions used are the **UNIFORM** distribution, **TRIANGULAR** distribution, and the **BETA** distribution (Figure 6.2). The triangular and beta distributions require three estimates: most pessimistic (minimum), most likely (mode) and most optimistic (maximum), whereas the uniform distribution requires two estimates: minimum and maximum estimates.

A risk analysis provides a comprehensive profile of the potential variability in the key performance criteria. However, in most cases, a sensitivity analysis using the most pessimistic scenarios is capable of providing sufficient information about the riskiness of a project or group of projects.

In the following section we will subject the project performance criteria for the Bintuli Wastewater Treatment Project to risk analysis. We begin by making qualitative assessments of the financial, economic and environmental risks associated with the project. We then proceed to use @RISK to estimate the impacts of uncertainty on the project performance criteria.

Figure 6.2 Probability distribution functions

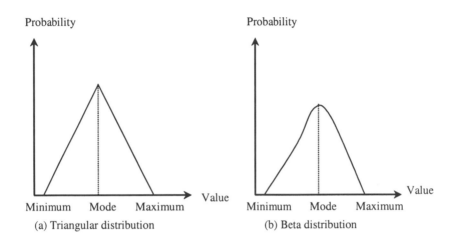

(a) Triangular distribution   (b) Beta distribution

6.10.1   *Conducting risk analysis for Case Study 1, Bintuli Wastewater Treatment Project*

The major sources of risk in this project are financial, economic and environmental. Each of these is briefly discussed.

Financial Risks

The major sources of financial risk for the project are the possibilities of cost overruns and the ability of the town council to raise the projected revenues. Shortfalls in revenue will increase the financial risk associated with the project. However, in view of the country's impressive economic growth in recent years, this particular risk is assessed to be negligible.

Economic Risks

Given that this is an environmental improvement project, there are no significant economic costs. Most of the expected benefits are in the form of cost savings. There is a risk that the projected economic benefits may not materialise if, for example, the water quality is below the expected level. In that event, the ability of users to pay would be reduced and the benefit estimates would be overstated. However, this risk is assessed to be minimal because the chosen treatment system has been found to be reliable.

Environmental Risks

There is a small but real risk of flooding of at least part of the project site. According to flood records, the last flood (classified as a 1 in a 100 year flood) occurred in 1963. Some of the water flowed into the old river channel, but drained underground quickly and none escaped back to the river. Although it is claimed by local officials that policies and standards for industrial waste disposal are already in place, there is a risk that lower treatment standards will be adopted in some factories and that additional pollutants will enter the waste stream.

Conducting the Risk Analysis

The qualitative risk assessment suggests that the major sources of risks are cost overruns and the possibility of the economic benefits not materialising. The following subjective assessment of the probabilities were made, using a triangular distribution. For example, the most optimistic assessment of the probablity of economic benefits was 5 percent above the project estimates, and the most pessimistic was 30 percent below the project estimates (Table 6.5).

Table 6.5 Subjective risk assessment, Bintuli Wastewater Treatment Project

Type of risk	Most pessimistic	Most likely	Most optimistic
Financial risk (cost overruns)	+30%	+5%	−1%
Economic risk	−30%	0%	5%

Note that, unlike sensitivity analysis, risk analysis considers the simultaneous impacts of the changes in risk on the project performance criteria. Figure 6.3 presents the distribution of the net incremental economic benefits generated with @RISK. The mean, maximum and minimum net economic benefits (in present value terms) over the life of the project are displayed. These figures were generated using the triangular distributions indicated in Table 6.5. It can be seen that the greatest variation in net economic benefits (and hence risk) is observed in the first four years.

Figure 6.3 Distribution of net incremental economic benefits, Bintuli Wastewater Treatment Project

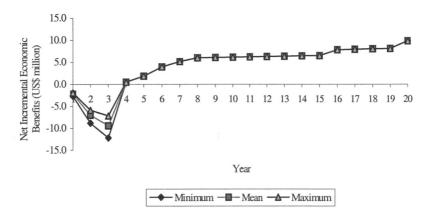

Table 6.6 presents the results of the Monte Carlo simulations for the NPV and IRR. The results indicate that even in the worst case scenario of 30 percent cost overruns and a 30 percent decline in expected net economic benefits, the project still returns a positive NPV and an IRR above the opportunity cost of capital of 12 percent. These results, together with those of the sensitivity analysis, suggest that the estimated project performance criteria are robust.

Table 6.6 Results of Monte Carlo simulations, Bintuli Wastewater Treatment Project

Project performance criterion	Minimum	Mean	Maximum	Standard deviation
NPV @12 percent	8.42	10.62	12.52	0.92
IRR (%)	17.5	19.6	21.7	0.9

Recommendations

Water pollution in the town of Bintuli is a serious problem. Currently about 70 percent of untreated effluent is dumped into nearby rivers, creating a health hazard. Implementation of the project is considered a matter of urgency in order to protect the health of the community and to reduce the rate of environmental degradation.

The project would yield substantial economic benefits. The IRR is estimated at about 21 percent. The sensitivity and risk analyses suggest that the estimate is insensitive to large changes in the projected economic benefits and costs. It is recommended that the project go ahead subject to the institution of a monitoring program covering physical, chemical and ecological parameters upstream and downstream of the effluent discharge location. The hydraulic, biological and chemical parameters of the treatment processes should also be monitored.

6.11   Summary

Under the competitive market assumption, the price a person is willing to pay for a good is proportional to the satisfaction he or she obtains from consuming an additional unit. We saw in the previous chapter that this assumption may not hold in the case of environmental goods and services. Therefore, in valuing such items, either non-market valuation techniques have to be used or shadow prices have to be derived. It was explained that the concept of 'cost' in social CBA is based on the principle of opportunity cost. The objective of social CBA is to determine whether net social benefits, that is willingness-to-pay minus opportunity costs, is positive. The steps involved in conducting a social CBA were introduced. Discounted cash flow techniques were used to demonstrate the calculation of project performance criteria such as NPV and IRR. Net present value was presented as a preferred project performance criterion, but the IRR could be useful for ranking projects with positive NPVs. Finally, the procedures were illustrated using the Bintuli Wastewater Treatment Project.

Key Terms and Concepts

beta distribution
break-even analysis
break-even value
capital cost or outlay
cost-benefit analysis
constant prices
depreciation
diminishing marginal utility
diminishing value method
discounted cash flow
discounting
economic benefits
economic life
economic risk
environmental risk
financial risk
implicit costs
incremental approach
incremental net benefits
incremental net economic costs
intensity of preference
linear method

marginal benefit curve
marginal rate of substitution
marginal utility of income
net incremental benefit
operating costs
planning period or horizon
primary costs and benefits
real rate
residual (or salvage) values
risk
risk analysis
risk margin
risk-free rate
secondary costs and benefits
sensitivity analysis
shadow pricing
subjective risk assessment
switching or cross-over value
triangular distribution
uncertainty
uniform distribution
working capital

Review Questions

1. Explain the terms 'willingness-to-pay' and 'opportunity cost'.

2. Explain why the price of a good can be used to determine the value of a good, if markets are perfectly competitive. Provide three examples of situations where this rule may not apply. State your reasons.

3. Explain two approaches for calculating the residual value of an asset.

4. What is working capital? How is it accounted for in cost-benefit analysis?

5. Explain why it would be erroneous to use net benefits 'before' and 'after' to measure a project's worth.

6. State reasons in support of and against the use of high discount rates in CBA.

7. Explain the meaning of the terms 'risk analysis' and 'sensitivity analysis'. What is their purpose in cost-benefit analysis?

Exercises

The cost and revenue streams (in $'000s) for two projects are given as follows.

Year	Project A			Project B		
	Capital Cost	O & M Cost	Total Revenue	Capital Cost	O & M Cost	Total Revenue
1	500	-	-	1000	-	-
2	500	-	-	1000	-	-
3		200	200		100	400
4		200	200		100	400
5		200	400		100	600
6		200	400		100	600
7		200	600		100	800
8		200	600		100	800
9		200	800		100	1000
10		200	1000		100	1200

1. Preferably using a spreadsheet, compute the NPVs using a 10 percent discount rate. On the basis of your results, rank the projects in terms of profitability.

2. Calculate the internal rate of return using a spreadsheet. Based on your results, rank the projects in terms of profitability. Are your rankings the same as in (1) above?

3. Calculate the cost-benefit ratio. Based on your results, rank the projects in terms of profitability. Are your rankings the same as in (1) and (2) above?

4. Comment on the discrepancies in the results in (1), (2) and (3).

5. Refer to the above data. For Option B, analyse the sensitivity of the calculated IRR with respect to ±40% changes in total costs and total revenue. Comment on the sensitivity of the estimated IRR.

6. A local government is considering two options for regulating pollution. These are: (i) to allow pollution and clean up afterwards, and (ii) to prevent pollution. The discounted costs and benefits of the two options are as follows.

Option	Total cost ($ '000s)	Total benefits ($ '000s)	NPV ($ '000s)
1. Pollute and clean up	2500	150	2350
2. Prevent pollution	5500	300	5200

   a. On the basis of the NPVs which option would you recommend?

   b. If the risk assessment indicates that there is a possibility of irreversible damage to the environment, which option would you recommend?

References

Dasgupta, P., Sen, A., and Marglin. S. (1972). *Guidelines for Project Evaluation.* Project Formulation and Evaluation Series o. 2, United Nations Industrial Development Organisation, Vienna.

Department of Finance (1991). *Handbook of Cost-Benefit Analysis.* Australian Government Publishing Service, Canberra.

Goodin, R. (1982). Discounting Discounting. *Journal of Public Policy*, 2: 53-72.

Hertz, D.B. and Thomas, H. (1983). *Risk Analysis and Its Applications.* John Wiley and Sons, New York.

Krautkraemer, J.A. (1985). Optimal Growth, Resource Amenities, and the Preservation of Natural Environments. *Review of Economic Studies*, 52: 153-170.

Little, I.M.D. and Mirlees, J.A. (1974). *Project Appraisal and Planning for Developing Countries.* Heineman Educational Books, London.

Megill, R.E. (1985). *An Introduction to Risk Analysis*, PennWell Books, Tulsa, Oklahoma, 2nd Edition.

Morgan, M. G. and Henrion, M. (1990). *Uncertainty.* Cambridge University Press.

Palisade Corporation (1997). *@RISK for Windows Version 3.5e.* Newfield, New York.

Pearce, D. (1983). *Cost-Benefit Analysis.* Macmillan, Basingstoke, 2nd Edition.

Pearce, D. (1987). Foundations of an Ecological Economics. *Ecological Modelling*, 38: 9-18.

Pearce, D. and Markandya, D. (1990). Environmental Sustainability and Cost-Benefit Analysis. *Environment and Planning*, 22: 1259-1266.

Schweitzer, M., Trossman, E. and Lawson, G.H. (1986*). Break-Even Analysis: Basic Model, Variants, and Extensions.* John Wiley, Chichester.

Squire, L. and van der Tak, H.G. (1975). *Economic Analysis of Projects.* International Bank for Reconstruction and Development and The Johns Hopkins University Press, Baltimore.

Appendix 6.1 An Application of Cost-Benefit Analysis (with Risk Analysis) to a Climate Change Abatement Strategy[43]

Introduction

Human economic activities over the last 200 years have increased the amounts of greenhouse gases in the atmosphere. Carbon dioxide, methane, nitrous oxide, and cholorofluorocarbons are the most important. The increased concentrations in the atmosphere of these gases have resulted in the so-called enhanced greenhouse effect. The enhanced greenhouse effect is caused by these gases acting as a shield and trapping a much larger than normal amount of outgoing solar energy in the lower atmosphere. The cooling capacity of the earth is reduced as the temperatures are increased, resulting in global warming. The effects of unabated global warming include rise in sea levels, increased incidence and severity of droughts, loss of ecosystems and animal species, and reduced outputs of agriculture especially in tropical areas.

Because of the widespread publicity about the negative effects of international warming, most countries have committed themselves to reducing emissions of greenhouse gases with the aim of stabilising the concentration of these gases in the atmosphere. Major agreements on the reduction of emissions of greenhouse gases were achieved at conferences in Montreal and Toronto. These agreements require industrialised countries to initially reduce their annual emissions of greenhouse gases to 1990 levels. Reduction of greenhouse gases involves costs to the economy of a country in the short term. However the gains to be realised later, if all countries undertake reductions of greenhouse gases, may outweigh the costs. The next two sections evaluate a major climate change abatement project from two perspectives: a traditional CBA and CBA incorporating risk analysis.

---

[43] This appendix is adapted from a publication entitled, 'An Introductory Discussion of Cost Benefit Analysis Applied to Climate Change Issues', Working Paper 9601, May 1996, Graduate School of the Environment, Macquarie University, Sydney, Australia, written by Kwabena A. Anaman.

Analysis of the Climate Change Abatement Project Using CBA
Without Incorporating Risk

Our hypothetical country is called the Republic of Olympia. The main effects of climate change as a result of continued emission of greenhouse gases are assumed to be more severe, frequent and longer droughts. Currently, in the Republic of Olympia, climate change effects are not noticeable and severe droughts occur on average once every 10 years as have been occurring over the last 150 years since the advent of modern records keeping of meteorological data and severe events. Without a successful global action by most countries to reduce their emissions of greenhouse gases, the Republic of Olympia will begin to experience the impacts of climatic change in 20 years time, starting in year 2021. Thus, there are two clear strategies: (i) the 'do nothing' strategy whereby the status quo of continued emission of greenhouse gases around the world is continued unabated and, (ii) the reduction of greenhouse gas production from year 2000 onwards to the 1990 emissions levels. The Republic of Olympia will only reduce its emissions if most other countries participate in reducing their emissions under various international agreements. It will be futile for the Republic of Olympia to do otherwise since it is a small producer of greenhouse gases; any unilateral reduction will have no effect on global warming while it suffers economic losses by reducing national output. It is assumed that year 2000 is the starting point for reduction of greenhouse gases emissions. The reduction will cost the Republic of Olympia 10 million dollars per annum in real 2000 constant prices.

With the do-nothing strategy, it is assumed for simplification that severe drought as a result of climate change would occur, on average, once every 10 years hypothetically in years 1, 11 and 21 in the Republic of Olympia. But after year 20, climate change effects become more pronounced because of the doubling of greenhouse gases in the atmosphere with severe droughts then occurring every 5 years starting in year 21 (year 2021), lasting for two years and costing the Republic of Olympia $200 million, double the $100 million cost for a severe drought before the year 21. Thus, after the climatic change impacts begin to be felt, severe droughts would occur in years 21 and 22, 26 and 27, 31 and 32, 36 and 37, 41and 42 with each year's losses being valued at $200 million. The analysis could even be extended to allow for the year a severe drought occurs to be randomly selected.

With the climate change abatement strategy approved and implemented by all signatory countries, The Republic of Olympia undertakes a climate change reduction project costing $10 million each year from year 1 to year 20 (i.e. 2001 to 2020). Given the widespread implementation of reduction of emissions, climate change impacts are expected to stabilise. Hence under this scenario, it is assumed that the existing natural climatic variability would be maintained. The predicted climate change that would have occurred starting from year 21 would be avoided by the stabilisation of greenhouse gases at 1990 production levels around the world. Hence, severe drought would occur as usual once every 10 years in the Republic of Olympia in years 1, 11, 21, 31 and 41 costing the economy $100 million.

Traditional CBA is used to analyse the economic viability of the climate change abatement strategy undertaken by the Republic of Olympia. Columns 2 and 3 of Appendix Table 6.1 summarise the costs associated with the do-nothing strategy and climate impact abatement project, respectively. It is assumed that computable general equilibrium models have been used to derive the economic losses of both strategies. Loss of resources such as plants and animal species due to possible climatic change is measured by total economic value. In this case, the benefits of the climate abatement strategy can be defined as the avoided costs of doing nothing.

Appendix Table 6.1 indicates that the NPV of the climate change abatement project is $177.5 million and IRR is 11.05%. The climate change abatement strategy is therefore economically viable. However, these benefits are contingent on all other countries undertaking emissions reduction programmes such that the natural climatic variability is maintained and not worsened.

Incorporating Risk into the Cost-Benefit Analysis of the Climate Change Abatement Project

In the risk analysis approach, the procedure is identical to the traditional CBA except that, rather than assuming that the cost of severe drought due to climatic change for the do-nothing strategy is $200 million, a probability distribution is used to express this outcome. Due to the considerable uncertainty regarding the impacts of the climate change, it is assumed that the minimum value of this outcome is $100 million which is identical to current costs for severe drought. The most likely value is $200 million and the maximum possible value is $600 million. It is assumed that economists working together with climate change impacts researchers have arrived at these figures.

Several probability distributions are available for expressing this uncertain outcome. However the BETA probability distribution function, BETA(α_1, α_2) was chosen because of its versatility to accommodate many possible shapes of uncertain outcomes. For the BETA probability distribution, the mean of a stochastic or uncertain variable is approximated as follows:

Mean   =   [Minimum Value + (4 * Modal Value) + Optimistic Value]/6

The corresponding standard deviation of the stochastic variable for the BETA probability distribution is approximated as [Maximum Value—Minimum Value]/6. The risk analysis was performed using the Lotus 1-2-3 Add-In @ RISK software Version 2.01 program developed by Palisade Corporation (1992). This program allows the economic analyst to explicitly include the uncertainty in a variable with the desired probability distribution selected from a list of about 30 probability distribution functions. The stochastic variable is expressed as cells in the Lotus 1-2-3 spreadsheet as BETA probability distribution rather than as single most likely or certain figures as is done in the 'traditional' deterministic economic analysis reported in the earlier section. The mode of a general BETA probability function is $(\alpha_1-1)/(\alpha_1+\alpha_2-2)$; the mean is denoted as $(\alpha_1)/(\alpha_1+\alpha_2)$, where α_1 and α_2 are the shape parameters which are greater than zero.

Specific shape parameters have to be estimated first before running the risk program. For this problem these specific shape parameters are determined as follows. The minimum, mode and maximum annual cost to the Republic of Olympia of severe drought under unabated climatic change are $100, $200 and $600 million respectively. The range is therefore $500 million.

Estimated mean = [Minimum Value + (4 * Modal Value) + Optimistic Value]/6 = (100 +(4 * 200) + 600)/6 = 1500/6 = 250.

Mode = $(\alpha_1-1)/(\alpha_1+\alpha_2-2)$ = (Mode value—Minimum value)/Range = (100)/500 = 0.20.

Mean = $(\alpha_1)/(\alpha_1+\alpha_2)$ = (Mean value—Minimum value)/Range = 150/500 = 0.30.

Hence, α_1 and α_2, the shape parameters, are 1.8 and 5.2, respectively, meaning that the function is skewed to the right.

The @RISK command for the BETA probability function is therefore written as: $100 + 500 * @BETA(1.8, 5.2)$.

Using the estimated BETA probability function, 100 and 500 iterations or simulation runs (sample size) are performed to determine the mean and standard deviation of the NPV and IRR for the climate abatement project. The results are presented in Appendix Table 6.2. They indicate that there are no major differences between the NPV and IRR of the project derived from either the 100 or 500 simulation runs. However the NPV and IRR values obtained from the risk analysis are different from those obtained from the traditional deterministic analysis reported earlier.

Appendix Table 6.1 Cost-benefit analysis of a climate change abatement project

Year number (Actual year) (A)	Cost of do nothing strategy ($ million) (without project) (B)	Cost of climate change abatement strategy ($million) (with project) (C)	Incremental net benefits ($ million) (D = B–C)
1 (2001)	100	110	−10
2 (2002)	0	10	−10
3 (2003)	0	10	−10
4 (2004)	0	10	−10
5 (2005)	0	10	−10
6 (2006)	0	10	−10
7 (2007)	0	10	−10
8 (2008)	0	10	−10
9 (2009)	0	10	−10
10 (2010)	0	10	−10
11 (2011)	100	110	−10
12 (2012)	0	10	−10
13 (2013)	0	10	−10
14 (2014)	0	10	−10
15 (2015)	0	10	−10
16 (2016)	0	10	−10
17 (2017)	0	10	−10
18 (2018)	0	10	−10
19 (2019)	0	10	−10

Appendix Table 6.1 (*Continued*)

20 (2020)	0	10	−10
21 (2021)	200	100	100
22 (2022)	200	0	200
23 (2023)	0	0	0
24 (2024)	0	0	0
25 (2025)	0	0	0
26 (2026)	200	0	200
27 (2027)	200	0	200
28 (2028)	0	0	0
29 (2029)	0	0	0
30 (2030)	0	0	0
31 (2031)	200	100	100
32 (2032)	200	0	200
33 (2033)	0	0	0
34 (2034)	0	0	0
35 (2035)	0	0	0
36 (2036)	200	0	200
37 (2037)	200	0	200
38 (2038)	0	0	0
39 (2039)	0	0	0
40 (2040	0	0	0
41 (2041)	200	100	100
42 (2042)	200	0	200
43 (2043)	0	0	0
44 (2044)	0	0	0
45 (2045)	0	0	0
		NPV @ 6% =	$177.25
		IRR =	11%

Notes:

(i)   It is assumed that severe drought as a result of climate change occurs with the do-nothing strategy once every 10 years in years 1, 11 and 21. But after year 20, climate change effects become pronounced because of doubling of greenhouse gases with severe droughts occurring every 5 years starting in year 21 and returning in years 26, 31, 36 and 41.

(ii)   It is assumed with the climate abatement strategy, natural climatic variability is maintained. The predicted climate change that would have occurred starting from year 21 is avoided by the stabilisation of greenhouse gases at 1990 production levels by all signatory countries around the world. Hence severe drought occurs once every 10 years in the Republic of Olympia for years 1, 11, 21, 31 and 41.

Appendix Table 6.2 Estimated NPV and IRR of the climate change abatement project

Item	Sample size	
	100	500
Mean NPV ($ million)	227	227
Standard deviation of NPV($ million)	48	45
Range of NPV($ million)	240	295
Mean IRR (%)	11.9	11.9
Standard deviation of IRR (%)	0.9	0.8
Range of IRR (%)	4.3	4.6

Note:
The NPV figures are rounded off to the nearest million dollars while the IRR figures are rounded off to the nearest decimal point.

References

Palisade Corporation (1992). @RISK: Risk Analysis and Simulation Add-In for Lotus 1-2-3 Version 2.01 Users' Guide, New York.

7.   Cost-Effectiveness Analysis, Impact Analysis, and Stakeholder Analysis

Objectives

After studying this chapter you should be in a position to:

❑ distinguish between cost-benefit analysis and cost effectiveness analysis

❑ conduct a cost-effectiveness analysis

❑ explain the components of an impact analysis

❑ explain the procedure of a damage assessment; and

❑ explain the steps in stakeholder analysis

7.1   Introduction

In the previous chapter, we introduced cost-benefit analysis as a framework for evaluating and ranking one or more investment alternatives. Although CBA is a useful tool to assist decision-making, it may not be a suitable approach in all situations. When it is not possible to value a project's major benefits in dollar terms, or when two projects have similar economic benefits, then a **cost effectiveness analysis** (CEA) may be used. For example, if the decision problem is to choose between building two hospitals, a CEA would be appropriate since the social benefits in either case would be similar.

Both CBA and CEA are based on the principle of economic efficiency and therefore do not consider equity or distributional issues. That is, a project is deemed to be socially desirable if the gainers can potentially compensate the losers. They both do not deal with the issue of who the losers are or how they should be compensated. Many projects around the world have resulted in failure because insuffient attention has been paid to equity and distributional issues. In this chapter, we consider **impact analysis** (or environmental impact analysis) and the relatively new area of **stakeholder analysis**. These two methods must be seen as being complementary to, and not a substitute for, CBA and CEA. They enable decision-makers to devise policies to mitigate any adverse impacts of a project on groups or individuals.

7.2   Cost-Effectiveness Analysis

Cost effectiveness analysis looks only at financial costs. A CEA takes the objective as given, and then works out the costs of the alternative ways of achieving that objective. The decision on whether to use CEA instead of CBA will depend on a number of factors including the following:

- the size and complexity of the project;
- the extent to which there are quantifiable benefits; and
- the extent to which the benefits can be valued in monetary terms.

For large-scale projects CBA is the preferred approach because it enables the major items of costs and benefits to be identified and valued and DCF performance criteria to be computed. However, in cases where the major benefits cannot be quantified in dollar terms, CEA is the preferred approach. CEA is also appropriate in a case where the choice is between, say, two wastewater treatment options with the same outputs or service levels but the difference is in, say, location. Most of the foregoing discussion on CBA applies generally to CEA. Unlike CBA, CEA does not have absolute criteria by which to judge the economic viability of projects. CEA is therefore not recommended when a decision about the level of output or service to be provided is at issue.

7.2.1 Examples of Cost Effectiveness Analysis

Examples of situations in which CEA could be used include the following:

1. Given a desirable pollution abatement standard, what will be the least cost, out of various alternatives, of achieving the standard?
2. Can buying up all the property rights in a flood plain and moving people out by constructing dykes save the same number of lives more cheaply?
3. Given two parks with similar recreation benefits, which should be developed? Park A requires extensive filling and flood control and Park B involves buying warehouse sites.
4. Choosing between alternative ways of constructing a town's water supply system.

7.2.2 Conducting a Cost-Effectiveness Analysis

The steps involved in carrying out a CEA are similar those for a CBA. They are:

1. Project definition
2. Choice of method of analysis
3. Identification and valuation of costs and benefits
4. Discounted cash flow analysis
5. Calculation of measures of effectiveness, and
6. Sensitivity analysis

Project definition

The issues here are the same as discussed for a CBA. That is, there is need to outline the project objective or problem to be solved, describe the target population, discount rate, project life and other useful parameters.

Choice of method of analysis

Careful consideration needs to be given to the method of analysis to use in a given situation. As already indicated above, CBA is the preferred approach in large and complex projects with significant social and environmental ramifications. However, in cases where two or more options have similar service levels and the major economic benefits cannot be valued in monetary terms, a CEA is appropriate.

Identification and valuation of costs

All the costs of the project must be identified and valued. Caution must be exercised in dealing with secondary (or indirect) costs. As far as possible, only direct costs must be used.

Discounted cash flow analysis

The stream of costs must be discounted to arrive at present values as described in the earlier sections.

Calculation of measures of effectiveness

An appropriate measure of effectiveness must be identified. As a guide, the measure must be as close as possible to the objective of the project activities. For example, for a wastewater project the obvious choice would be 'cost per Ml' or 'cost per Ml per annum'.

Sensitivity analysis

As in a CBA, a sensitivity analysis will be required in a CEA. To assist the choice between options, it would be necessary to find out how sensitive the cost-effectiveness measures are to changes in project parameters such as discount rate and planning horizon.

7.2.3   *Example of a CEA Application: The Cooloombah Wastewater Project*

This section presents an example of a CEA application. The intention is to highlight and provide practical examples of the essential features of a CEA. In this respect, this example must not be taken as an exhaustive review of CEA methodology. We begin with a brief description of the project background.

Background

Cooloombah is a small city of 40,000 people located to the northeast of Brisbane. In line with the general trends in the state of Queensland, the population is expected to triple by the year 2020. To meet this increase in the city's population, Council is considering two wastewater treatment options. Option 1 involves providing additional capacity at the existing facility to cater for the expected increased population. Option 2 is to abandon the existing facility and to construct a new one to cater for the whole catchment. Both options are capable of treating an additional 20 Ml/d of treated wastewater for discharge into the neighbouring creek. The estimated costs are listed in Table 7.1.

Project definition

The project's objective is to augment the wastewater treatment capacity in the project area. This goal can be achieved in two ways: to expand capacity at the existing plant (Option A) or to build a new plant (Option B).

Table 7.1 Cost estimates for Cooloombah Wastewater Project

Cost category	Augment existing facility	Build new facility
Capital cost ($ million):		
Investigations	0.03	0.05
Surveys	0.02	0.03
Engineering	1.50	1.00
Plant and equipment	48.45	40.00
Land acquisition	-	2.00
Total	50.00	43.08
Operating cost ($ million per annum):		
Labour	2.00	2.00
Electricity	0.30	0.30
Chemicals	0.40	0.20
Repairs and maintenance	0.50	0.40
Sludge management and odour control	0.30	0.10
Insurance	0.10	0.08
Administrative overheads	0.20	0.10
Total	3.80	3.18

Choice of method of analysis

The two options under consideration have similar output levels. Apart from improvement in morbidity and mortality rates as a result of treating the wastewater, there are no obvious tangible economic benefits because there are no plans for reusing the treated water.

Identification and valuation of costs and benefits

The benefits of each option include:

- revenue from charges for wastewater treatment, and
- improvement in morbidity and mortality rates as a result of treating the wastewater.

The major costs consist of the following:

- capital costs associated with surveying and engineering, as well as purchasing and installing plant equipment, and
- plant operating costs.

Estimates of costs for each option are as follows:

Option A:
The capital costs of augmenting the current facilities are estimated at $50 million and will be phased evenly over the construction period of two years. Operating costs are estimated at $3.8 million per annum and will increase at a rate of 3 percent per annum above the projected rate of inflation. Operating costs are therefore projected to increase from $3.8 million at the start of operation to $6.1 million at the end of the planning horizon (Table 7.2).

Option B:
The capital costs of building a new wastewater treatment plant are expected to amount to $43 million over a period of two years. The operating costs will be $3.2 million per annum, increasing at 3 percent per annum above the expected rate of inflation. The operating costs will therefore be $3.2 million in the first year of plant operation, increasing to $5.14 million at the end of the project (Table 7.2).

Discounted cash flow analysis

The discounted cash flow analysis (Table 7.2) indicates that the present values of the total costs are $80.63 million for Option A and $68.70 million for Option B, using a real discount rate of 8 percent.

Calculation of measures of effectiveness

A useful measure of effectiveness is the cost per unit volume of treated water. The discounted volume of water to be treated over the project period is 55.69 million kilolitres. The cost-effectiveness of the two options are therefore $1.45/Kl for Option A and $1.23/Kl for Option B. From these figures, it would appear that construction of a new wastewater plant is more cost effective than augmenting the current facility.

Recommendation

It is recommended that Council consider construction of a new wastewater plant.

7.3   Impact Analysis

As already indicated earlier, a CBA (or CEA) has nothing to say about who the losers of a project are what is the magnitude of their losses. An impact analysis (IA), or environmental impact analysis (EIA), complements a CBA by assessing a project's impact on the environment, social groups or individuals, as well as the economic impacts.

Although there are different approaches to conducting an IA or EIA, at least three major components may be identified: an **environmental impact assessment**, an **economic impact assessment** and a **social/cultural impact assessment**.

The environmental impact assessment considers the project's impacts on the flora and fauna, land and soils, noise and pollution, and so on. These days, many countries (both developed and developing) have legislation that require an EIA when substantial public and private programs and projects are under consideration. The social/cultural impact assessment examines the impact of the project on individual groups within the community by identifying who they are and how better off or worse off they are if the project goes ahead.

Table 7.2 Discounted cash flow table for Cooloombah Wastewater Project
($ millions)

Year	Option 1 Capital Costs	Option 1 Operating Costs	Total Costs Option 1	Option 2 Capital Costs	Option 2 Operating Costs	Total Costs Option 2
1	25.00		25.00	21.50		21.50
2	25.00		25.00	21.50		21.50
3		3.80	3.80		3.20	3.20
4		3.91	3.91		3.30	3.30
5		4.03	4.03		3.39	3.39
6		4.15	4.15		3.50	3.50
7		4.28	4.28		3.60	3.60
8		4.41	4.41		3.71	3.71
9		4.54	4.54		3.82	3.82
10		4.67	4.67		3.94	3.94
11		4.81	4.81		4.05	4.05
12		4.96	4.96		4.18	4.18
13		5.11	5.11		4.30	4.30
14		5.26	5.26		4.43	4.43
15		5.42	5.42		4.56	4.56
16		5.58	5.58		4.70	4.70
17		5.75	5.75		4.84	4.84
18		5.92	5.92		4.99	4.99
19		6.10	6.10		5.14	5.14
		PV(costs, $m.) =	80.63		PV(costs, $m.) =	68.70
		Cost/Kl ($) =	1.45		Cost/Kl ($) =	1.23

The analysis should, as far as possible, indicate the size and nature of the gains and losses. A simple but practical way of reporting the results of the social impact assessment is by means of a matrix, depicting the costs and benefits on one axis and the relevant socio-economic groupings on the other. The economic impact assessment evaluates the impact of the project on the region (or state), on local communities and on industry sectors. The economic impact assessment can be carried out, at an elementary level, by describing the impacts and providing estimates of how many jobs will be created during the construction and operation phases of the project. A more sophisticated analysis can be carried out using either an input-output or a computable general equilibrium model. These models can give an indication of the 'economy-wide' impacts of the projects. For example, if a project increases output of a good or service, the models can provide an indication of the simultaneous impacts on different sectors of the economy, including effects on government revenue and households. However, it must be emphasised that these models are not appropriate for project selection because they do not consider the alternative uses of the resources consumed by a project.

7.3.1 Example of an Impact Analysis in Case Study 1, Bintuli Wastewater Treatment Project

The impacts of the project can be grouped under three main headings: environmental, economic and social. A brief assessment of these impacts is provided below.

Environmental impacts

The environmental impacts will occur during the construction and operation phases. These include impacts on the water quality, land and soils, flora and fauna, and noise and air pollution.

Water quality

The impact on water quality will be positive because the treatment process will remove nutrients from the wastewater and will also disinfect the wastewater by means of chlorination.

Flora and fauna

There are no significant numbers of wildlife in the area. It is not expected that there will be significant adverse impacts on the flora and fauna. On the contrary, improved river water quality will benefit flora and fauna over a large area.

Noise and air pollution

The site is isolated from settlements. Although there will be substantial noise and air pollution from equipment during construction of the plant including traffic, this would be transient and not significant. The existing roads are capable of accommodating the (relatively insignificant) levels of additional traffic.

Economic impacts

It is expected that construction of the plant will provide employment opportunities, especially during the construction period. It is estimated that 150 jobs will be created when the plant is completed.

Social impacts

The social impacts pertain to the displacement of people, as well as gender and cultural issues.

Displacement

The proposed site for the project is not currently populated. The site is not being used for any particular purpose and therefore there will be no displacement impact.

Gender

It is expected that the construction of the plant will not adversely affect women in the area. It is uncertain whether women in the area would share in the employment opportunities to be created during the construction and operation of the plant. There is little information about the education and skills level of women in the area.

Cultural

The site for the proposed plant has little cultural significance to the population. It is unlikely that there will be any important cultural impact from the project. There will also be no impacts on farmers, river users and fishermen.

7.4   Damage Assessment

The Comprehensive Environmental Response, Compensation and Liability Act was passed in the U.S. in 1980 authorising governments at various levels to act as trustees for publicly owned natural resources and to sue people responsible for environmental damage. This law, coupled with the 1989 Exxon Valdez oil disaster in Alaska, has given rise to a new type of environmental analysis known as **damage assessment**. The objective of damage assessment is to estimate the monetary value of environmental damage caused. This information could be used by courts to determine the liability of those responsible and to compensate the victims. For example, Exxon has paid US$1 billion for cleanup, and pending appeals, is liable to pay an additional US$5 billion in punitive damages.

In the case of damage assessment in the U.S., the Department of Interior (DOI) has determined that damages should be equal to the lesser of: (1) the lost value of the resource, or (2) the value of restoring the resource to its former state.

The lost economic values from environmental damages include: loss of income resulting from, say, damage to fishing grounds; loss of recreational opportunities (e.g. camping, fishing and hiking); pollution of water resources; and air pollution. The methods discussed in Chapter 5 could be used to determine some of these economic values. For example, in the case of the Exxon Valdez oil spill, the amount non-users were willing to pay to avoid the damage actually incurred was estimated to be US$2.8 billion, which is equivalent to US$31 per U.S. household (O'Doherty, 1994)

7.5   Stakeholder Analysis

The concept of stakeholder analysis (SA) originated in the area of business management.[44] In the early 1980s, business management strategy began to take a much broader view beyond the issue of making profit from the manufacture and distribution of goods and services. It was felt at the time that there was a need to take into account the concerns of not only direct stakeholders (i.e., those who directly buy or supply products to a firm) but also indirect stakeholders. Since the early 1990s, SA has been adopted for use in the area of natural resource management. Many projects and policies

---

[44] See, for example, work by Mitroff (1983) and Freeman (1984).

have failed in the past due to the failure of the planners to directly involve those who the project will affect.

Grimble and Chan (1995:114) define SA as 'an approach and procedure for gaining an understanding of a system by means of identifying the key actors or stakeholders in the system, and assessing their respective interests in the system'. 'Stakeholders', in this definition, include governments and non-governmental organisations, individuals, community groups, and firms who can potentially affect or be affected by a proposed policy or project. The main aim of SA is to take account of distributional and equity issues in the design of policies or projects.

The justification for an approach such as SA is on the grounds that conventional approaches such as CBA and CEA, particularly in the area of natural resource management, do not consider the distribution of costs and benefits among the stakeholders. Since stakeholders perceive environmental problems differently, they may pursue solutions that are different from those of the project, leading to non-cooperation with the project. Although SA is useful in development planning as a whole, it is particularly useful for natural resource management due to a number of reasons. Among these are the common property nature of some natural resources; the multiple use nature of natural resources and the varying interests held by different types of stakeholders. In such situations stakeholder analysis provides a consistent framework for evaluating and incorporating the various interests in the decision-making process.

7.5.1   Steps in Stakeholder Analysis

Stakeholder analysis involves the following steps:

- identifying the objective(s) of the analysis;
- developing an understanding of the system and its decision makers;
- identifying the stakeholders;
- investigating stakeholder interests; and
- identifying patterns and contexts of interaction between stakeholders

Identifying the objective of the analysis

This first step of identifying and clarifying the purpose of the analysis is useful for other kinds of analysis, but more so in SA because there is a wide range of resource management problems. It is therefore necessary to determine the nature of the problem to be solved, the objectives of the analysis, the outputs of the analysis, the relevant decision-makers and the main target group. For example, in a dam project the problem to be addressed could be the fact that some groups would be worse off (e.g., suffer income losses). The objectives and outputs would therefore be to consider alternative project designs to minimise the impacts.

Developing an understanding of the system and its decision makers

Once the objective of the analysis has been firmly established, the next step is to gain an understanding of how the system operates and who the major players are. In the case of natural resource systems, it is important to understand the interactions between the local community and the environment. For example, if the problem were deforestation, it would be essential to understand the various uses such as logging, grazing, fruit harvesting and so on.

Identifying the stakeholders

Various approaches for identifying major stakeholders have been proposed in the management science literature. These include the following: the **reputational approach**, the **focal group** approach and the **demographic approach**.

The reputational approach involves seeking assistance from prominent individuals (e.g., the village chief) in the community for help to identify groups or individuals who may have a stake in the resource issue at hand. As the name suggests, the focal group approach involves first identifying a main group with interest in the issue, and then looking for other groups with links to this main group. Finally, the demographic approach attempts to identify stakeholders on the basis of demographic characteristics such as age, gender, religion and so on. After the stakeholders have been identified, a list of stakeholders is compiled and the information is verified. Verification of the stakeholders can be carried out by conferring with other stakeholders.

Investigating stakeholder interests

Once the stakeholders have been identified, the next step is to investigate their interests. Table 7.3 provides an example of the various stakeholders of a forestry resource in a developing country and their interests. The table indicates that there is a range of contexts within which the various interests operate. Techniques such as Participatory Rural Appraisal (PRA) or Rapid Rural Appraisal (RRA) may be used to collect the necessary information (IIED, 1994). Information to be collected includes how stakeholders use and manage the given resource, the direct and indirect benefits, the system of property rights, stakeholder views about the resource use, actual and perceived costs and benefits, stakeholders' decision-making criteria, and other relevant information.

Table 7.3 An example of stakeholders of a forest resource

Institutional level	Example of stakeholder	Environmental interest
Global and international; wider society	International agencies; foreign governments; lobbies	Biodiversity conservation; climatic regulation; maintenance of resource base
National	National governments; macro planners; urban pressure groups; NGOs	Timber extraction; tourism development; resource and catchment protection
Regional	Forest departments; regional authorities; downstream communities	Forest productivity; water supply protection; soil depletion and siltation
Local off site	Downstream communities; logging companies and sawmills; local officials	Protected water supply; access to timber supply; conflict avoidance
Local onsite	Forest dwellers; forest fringe farmers; livestock keepers; cottage industry	Land for cultivation; timber and non-timber; forest products; access to grazing and fodder; cultural sites

Source: Grimble and Chan (1995).

Identifying patterns and contexts of interaction between stakeholders

This stage of the process builds on Steps 2 to 4 by conducting a more detailed analysis of the relationships between the different stakeholders. The aim of this activity is to shed more light on conflicts or cooperative action among the stakeholders. The expected outcome of this component of SA is an understanding of the divergent interests and factors that are likely to influence successful cooperation among groups.

Stakeholder analysis is a useful tool to complement more conventional methods such as CBA and CEA. It has the potential to generate information that can be used to improve the design projects to ensure successful implementation.

7.6   Summary

In this chapter, we have considered cost effectiveness analysis as an alternative form of comparing two or more investment alternatives. A CEA can be used when a project has significant economic benefits that cannot be valued in monetary terms. It is also used to choose among two or more projects or policies that have identical benefits. In such cases, the issue of interest is which one can be implemented at least cost.

A practical example of a CEA was outlined. Both CBA and CEA are based on principles of economic efficiency and do not therefore consider equity and distributional issues associated with a proposed policy or project. The concepts of impact analysis and stakeholder analysis were therefore introduced as additional tools to complement CBA or CEA. An impact analysis identifies a project's impact on the environment, social groups and individuals. It seeks to identify gainers and losers and the magnitude of the gains and losses. Stakeholder analysis is a similar process in the sense that it seeks to identify the stakeholders, that is, individuals, groups and organisations who are affected by or affect a given policy or project. Stakeholder analysis attempts to gain an understanding of the various interests associated with a resource and the interactions among the stakeholders. Information from both impact and stakeholder analysis can be used to improve the design of projects so as to minimise any adverse impacts and improve the chances of successful implementation.

Key Terms and Concepts

cost effectiveness analysis
damage assessment
demographic approach
economic impact assessment
environmental impact assessment
focal group approach
impact analysis

measure of effectiveness
participatory rural appraisal
rapid rural appraisal
reputational approach
social/cultural impact assessment
stakeholder analysis

Review Questions

1. What are the main differences between cost-effectiveness analysis and cost-benefit analysis?

2. Describe the steps in carrying out a cost-effectiveness analysis.

3. Discuss the reasons why a cost-benefit analysis alone may be an insufficient basis for selecting some types of projects. What kinds of additional analyses may required?

4. Outline the steps in impact analysis.

5. Outline the steps in stakeholder analysis.

Exercises

1. List the advantages and limitations of cost effectiveness analysis.

2. List the advantages and limitations of stakeholder analysis.

References

Freeman, R.E. (1984). *Strategic Management: A Stakeholder Approach*. Pitman, Boston, MA.

Grimble, R. and Chan, M-K. (1995). Stakeholder Analysis for Natural Resource Management in Developing Counties. *Natural Resources Forum*, 19(2): 113-124.

International Institute for Environment and Development, IIED (1994). *Rapid Rural Appraisal Note 20*. Sustainable Agriculture Progamme, IIED, London.

Mitroff, I. (1983). *Stakeholders of the Organisational Mind*. Jossey-Bass, Ca.

O'Doherty, R. (1994). Pricing Environmental Disasters. *Economic Review*, 12:1.

8.   Multi-Criteria Analysis

Objectives

After studying this chapter you should be in position to:

❑   explain when it is appropriate to use multi-criteria analysis;

❑   explain the steps in multi-criteria analysis; and

❑   explain the strengths and weaknesses of multi-criteria analysis.

8.1   Introduction

The non-market valuation techniques discussed earlier aim to value the environmental impacts of proposed policies or projects in dollar terms. These dollar estimates are then used as inputs to an analytical framework such as CBA or CEA to assist decision makers choose between competing alternatives. Although it is not necessarily made explicit, it is usually the case that projects seek to achieve a set of broader socioeconomic and environmental objectives such as increased income and preservation of the environment. CBA works well as a decision-making tool when costs and benefits can be valued in dollar terms. However, the fact remains that, despite improvements in non-market valuation techniques, many environmental impacts cannot be valued in monetary terms. This chapter introduces multi-criteria analysis (MCA) which is an approach for choosing from among a set of alternatives when there are multiple objectives. Because the objectives may be competing or conflicting, the trade-offs need to be made explicit.

The chapter proceeds as follows. The next section defines MCA and explains its conceptual basis. Section 3 outlines the steps involved in conducting MCA, while Section 4 discusses the limitations of the approach. The penultimate section presents a case study on an application of MCA, while the final section contains the summary.

8.2   The Multi-Criteria Analysis Approach

Multi-criteria analysis is also referred to in the literature as multi-objective decision making, multi-objective decision support system (MODSS), and multi-criteria decision aid. MCA may be described as a framework to assist decision makers choose between alternative policies and projects in situations where there are multiple objectives. In contrast to the methods discussed previously, MCA incorporates costs and benefits that cannot be valued in dollar terms. As a matter of fact, CBA may be considered as a special case of MCA in which the alternatives are evaluated by performance criteria (e.g., NPV) that are measured in dollar terms.

Apart from not being based exclusively on money valuations, MCA differs from CBA (or CEA) in other respects. Both CBA and CEA are based on economic efficiency criteria (e.g., NPV $\geq$ 0), but MCA incorporates other types of criteria such as distributional, equity, ecological and so on.[45] Unlike CBA and CEA, MCA does not require only quantitative data. The approach can handle quantitative, qualitative and a combination of quantitative and qualitative data. In situations where efficiency is a major criterion, where trade-offs do not need to be identified explicitly and where there are few valuation problems, a CBA is preferable to an MCA. However, as indicated above, in many cases, the need to include social and ecological effects that cannot be valued in monetary terms makes MCA a more pragmatic alternative. According to van Pelt (1991), the choice between CBA and MCA boils down to the trade-off between methodological and empirical considerations.

To explain the conceptual basis of MCA, let us consider a potential project that has conflicting objectives: biodiversity protection (y_1) and development (y_2). In this case, development entails a loss of biodiversity. Assume that the additional costs required to protect biodiversity are known

---

[45] Equity considerations may be modelled in CBA, to some extent, by adjusting prices according to predefined equity weights.

and that the level of development can be measured against an index of biodiversity loss (Figure 8.1). Suppose that alternative projects *a*, *b*, *c* and *d* have been identified as potential solutions posed by the first project. Point *c* is preferable to *d* because it implies a lower cost as well as a lower biodiversity loss.

Figure 8.1 The trade-off curve in multi-criteria analysis

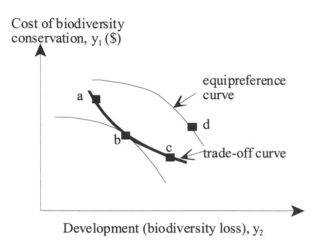

Cost of biodiversity conservation, y_1 ($)

Development (biodiversity loss), y_2

In this regard, alternative *d* may be discarded because it is inferior to (i.e., dominated by) *c*. However, a similar conclusion cannot be made about points *a* and *b* because they lie on the same **trade-off** curve. Point *c* is better in terms of having a lower cost but has a higher biodiversity loss compared to point *a*.

The process of MCA involves finding additional points such as *b* and *c* that form an optimal trade-off curve. Using the decision-maker's preferences (where the decision-maker could be a number of groups of stakeholders holding divergent opinions about resource management), the analyst constructs a family of trade-off and **equipreference** curves that describe how one objective is traded off against the other (Figure 8.1). The optimal choice is the one that yields the greatest utility. This is given by the point of tangency between a trade-off curve and the lowest equipreference curve, point *b*. The equipreference curves are often not known and therefore some other means must be used to narrow the set of feasible points on the trade-off curve. For example, using the safe minimum standard concept (see Chapter 11), the decision makers might define a maximum allowable level of biodiversity loss. Depending on the budget constraint, they may also define the maximum level of costs. Imposing such constraints enables the choice set to be narrowed down.

8.3   Steps in the MCA Process

Although the procedure for MCA is presented below as a series of successive steps, this is merely for ease of presentation. In practice, there is likely to be iteration between steps. The steps in MCA, are as follows:

- identifying the problem to be addressed
- identifying the alternatives
- identifying the criteria
- scoring the alternatives in relation to each criterion
- weighting the scores according to the weight or rank assigned to the criteria
- evaluating the alternatives
- sensitivity and risk analyses, and
- producing a ranking of alternatives on which to make a recommendation

These steps are represented schematically in Figure 8.2. It shows the two main sources of information flows (represented by broken arrows) during the MCA process. The first source is in the form of inputs from data sets. Data sets, typically, would include scientific, social and economic information about the problem to be addressed. It may be presented in the form of GIS, simulation models or expert systems that simulate the problem and alternative scenarios and provide predictive information to assist the identification of alternatives to address the perceived problem and the scoring of alternatives against the evaluation criteria.

The second source is information from identified stakeholders. MCA facilitates a participatory approach to decision-making. Information provided by stakeholders would be expected to include their input to identify alternatives and their acceptance of the appropriateness of the alternatives under consideration; the identification of the criteria to evaluate the alternatives as well as a ranking or weighting of the criteria.

Figure 8.2 Schematic representation of the steps in the MCA process

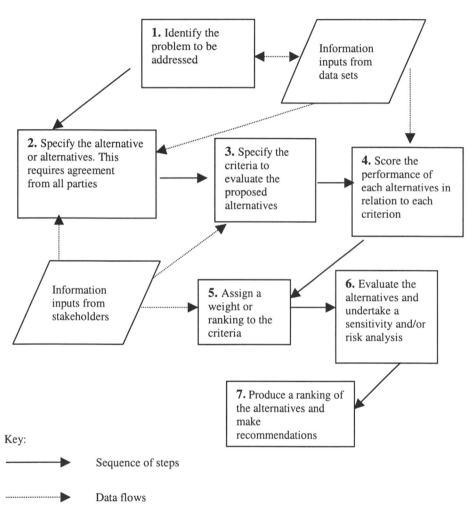

Key:

Source: Robinson (1999a).

8.3.1   Identifying the Problem to be Addressed

An initial, and preliminary step in the MCA process is the identification of the problem to be addressed. It is likely that this step would initiate data collection to determine the extent and severety of the problem. In the illustrative case here, the problem to be addressed would be identified as the sustainable use of forest resourses.

8.3.2   Identifying the Alternatives

Identification of the relevant set of alternatives is a crucial early step in MCA. Given that MCA is a process, the aim of this step is to seek the input of the stakeholders in deciding what alternatives should be considered. Obviously, there is a problem if there are either too few or too many alternatives. In the case of too many alternatives, preliminary screening can reduce the number of alternatives using some agreed criteria (e.g., minimum pollution levels or minimum economic returns). In this screening exercise, 'inferior' alternatives (i.e., those that are dominated by one or more alternatives) are dropped. In the case of forest resource use, there could be four possible alternatives which would include a 'do nothing' scenario:

1.   Ban all forms of logging;
2.   Ban logging in only old growth forest, allow logging in plantation forest;
3.   Permit logging of any type; and
4.   Do nothing.

8.3.3   Identifying the Criteria

As already indicated, one strength of MCA is its ability to accommodate decisions with multiple objectives. In addition, MCA can deal with hierarchical criteria. In this case, the criteria are grouped into sub-criteria, and if necessary, into sub-sub-criteria. For the forest management example, we can identify three main objectives (Figure 8.3), namely

1.   Maximise economic growth;
2.   Maximise environmental quality, and
3.   Maximise social benefits.

The criteria for the first objective could be the number of jobs created and the amount of revenue from the sale of logs. The criteria for the second objective could include the level of biodiversity, the area of forest conserved, the amount of soil erosion avoided and the level of water quality. Sub-criteria for water quality could include the pH level, amount of suspended particulate matter, and so on. Finally, the criteria for the third objective could be the value of recreation and tourism, and the level of poverty.

8.3.4 Scoring the Alternatives in Relation to the Criteria

Before scores are assigned to each alternative in relation to each criterion, the 'do nothing' alternative must be identified so that the alternatives can be evaluated as marginal or incremental to the 'do nothing' alternative.

Figure 8.3 Hierarchy of criteria for the forest management example

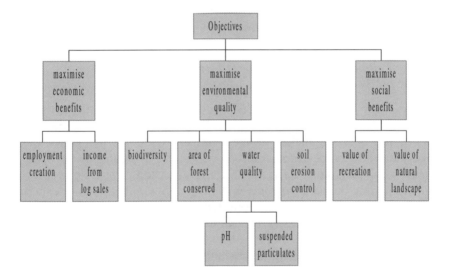

Scores are presented in an 'effects table'. An effects table, or matrix, displays the criteria in the rows and the alternatives in the columns. A score is provided for each alternative against each criterion in relation to the 'do nothing' alternative. As already indicated above, MCA has an advantage over CBA in that it is able to deal with both quantitative and qualitative criteria.

For example, for the employment creation criterion in the forest management case, the score for Alternative 4 (do nothing) could be no additional jobs, Alternative 1 (ban all logging) could be 3 jobs and that of Alternative 3 (permit all logging) could be 90 jobs (Table 8.1). For the income generation criterion, the score could be zero dollars per annum for both the 'do nothing' alternative as well as Alternative 1 and $35,000 per annum for Alternative 3. Note that some of the criteria are expressed in qualitative, or ordinal, form. For example, the score for the biodoversity is high under Alternative 1 and very low under Alternative 3. It is recommended that, as far as possible the criteria be 'standardised' (RAC, 1992). Standardisation involves reducing the criteria scores to a comparable basis. In the forestry management example, standardisation would mean ensuring that all the dollar scores are expressed on a scale such as between 0 and 10. Employment generation could be converted into a score, using, say, '10' for the highest estimated employment level, and '0' for the lowest.

8.3.5   Weighting the Scores According to the Weights Assigned to the Criteria

The next step in the MCA process involves 'prioritising' the criteria by assigning different rankings or weights. The weights can be assigned by the analyst, the decision maker or they can be based on the views of the stakeholders. In some cases, the criteria can be weighted by a panel of experts using techniques such as the **Delphi method**. The Delphi approach involves getting the panel to come up with weights in their area of expertise. Each person is then given the opportunity to re-evaluate their individual responses. Through successive rounds of re-evaluation, a consensus on the weights is then reached.

The weights can also be generated mathematically. For example, Janssen (1992) discusses three approaches to deriving weights, including one that assumes a linear relationship between pairs of criteria. Another approach is the 'analytical hierarchy process' in which weights are estimated based on pairwise comparisons (Saaty, 1980; Forman, 1990). The scores are then weighted according to the ranking or weights assigned to the criteria.

8.3.6 *Evaluating the Alternatives*

A number of approaches for evaluating the alternatives[46] have been identified for evaluating the feasible choices on the trade-off curve.

Table 8.1 Effects table for the resource management example

	Alternatives			
Criteria	1. Ban all logging	2. Ban logging in old growth forest	3. Permit all forms of logging	4. Do nothing
Employment creation (number of jobs)	3	30	90	0
Income ($'000 p.a.)	0	40	35	0
Biodiversity	high	low	very low	low
Forest conserved (ha)	10,000	6,000	1,000	5,000
Water quality (low to high)	high	medium	low	medium
Soil erosion (kg/ha/p.a.)	1	25	70	30
Recreation and tourism ($'000 p.a.)	500	300	10	250
Poverty level ($ change in per capita income)	−20	+5	+30	0

They include the following:

- aggregation techniques;
- lexicographic technique;
- graphical technique;
- consensus maximisation;
- concordance methods; and

---

[46] See, for example, Pearce and Turner (1990) and Janssen (1992).

- multi-attribute utility model.

Aggregation techniques

In the aggregation approach, the scores are aggregated over the range of criteria. The alternatives are then ranked according to the weighted sum of the standardised scores. The scores are defined as follows:

$$\sum_{j=1}^{n} w_j x_{ji} \qquad\qquad (8.1)$$

where: x_{ji} = performance of alternative i with respect to criterion j;
                w_j = weight assigned to criterion j; and
                n = number of criteria

Lexicographic technique

The lexicographic technique evaluates criteria by ranking them from the least important to the most important. This approach does not allow trade-offs among the criteria and is suitable for decisions in which the priorities are clear-cut. An example is a decision regarding concentrations of toxic chemicals in drinking water where the levels of acceptable risk can be clearly defined based on medical information.

Graphical technique

As the name suggests, the graphical approach involves plotting the alternatives in relation to benchmarks. These benchmarks could be, for example, the maximum scores for a given alternative. The main limitation of this approach is that the data must be in quantitative form (i.e., either ratio or interval data).

Consensus-maximising technique

In the consensus-maximising approach, individual preferences are added up to obtain group consensus preferences. This approach is useful when there are a range of stakeholders associated with the project in question. In this case, the criteria could be ranked on the basis of the preferences of different interest groups to obtain an aggregate index.

Concordance methods

Concordance methods are also referred to as ELECTRE methods (Roy, 1990). For each pairwise comparison, concordance and disconcordance indices are calculated. The concordance index is given by the weighted sum of the criteria for which the first alternative's score is better than that of the second. The disconcordance index is given by the largest difference in scores among those criteria for which the first alternative's score is below that of the second. Upon calculation of the indices, a ranking of the alternatives is then carried out.

Multi-attribute utility model

The multi-attribute utility approach is a generalisation of the aggregation method. In this approach a utility function is used in place of the scores, and the alternatives are ranked according to the following weighted sum:

$$\sum_{j=1}^{n} w_j U_j(x_{ji}) \qquad (8.2)$$

where U_j is a utility function; and all other variables are as defined previously.

8.3.7   Sensitivity and Risk Analyses

The assignment of weights and calculation of scores in the MCA process are carried out in an environment of imperfect information. In this regard, it is necessary to check how 'robust' the final solution is. This can be done by conducting a sensitivity analysis, which determines how the rankings change in response to changes in, say, the weights, the standardisation method and the evaluation method. Sensitivity analysis is commonly carried out to determine any change in the ranking of alternatives resulting from a change in the weights assigned to the criteria. Risk analysis may also be performed by using the Monte Carlo simulation approach (see Chapter 6) to generate probability distributions for the possible rankings. This particular type of risk analysis is more commonly conducted on data that are in a quantitative form. At the moment, there are no established simulation methods for qualitative (or ordinal) data.

8.3.8   Ranking the Alternatives and Making a Recommendation

The final step in MCA is to establish a ranking of the alternatives and to make a recommendation. The outcome from a MCA process is a prioritisation of alternative courses of action or projects. Depending on the number of alternatives and criteria, the process can generate a vast amount of information. Graphical methods have been shown to be an effective way of presenting the results for different alternatives (Janssen and van Herwijnen, 1991). Interactive computer packages (see, for example, DNR, 1999) are now available which enable the decision maker to view graphical outputs, as well as what happens if any of the key parameters or assumptions change

8.4   Limitations of MCA

The main advantage of MCA (or MODSS) over other analytical methods is that it enables a more realistic representation of the decision problem to be made, and in particular for the trade-offs to be made explicit. The interactive nature of the approach enables both the analyst and the decision maker, who could be a number of groups of stakeholders, to learn more about the problem. Although MCA is a structured approach which can incorporate techniques and data from other methods, it is flexible enough to allow the use of value judgement. As already indicated above, it is suitable for problems where dollar estimates of the effects are not readily available.

The MCA process, however, does have a number of inherent weaknesses. The credibility of the process will be determined ultimately by the quality of the data sets and the degree to which the stakeholders or interest groups have been involved in the process. There is a real danger that community preferences will be determined, not by the community, but by a single decision-maker, without consultation with the community. The criteria adopted to evaluate alternatives need to provide a balanced evaluation. This means that a tendency to include a number of economic criteria and only one social and one environmental criterion, needs to be avoided.

There is a wide variety of evaluation methods in MCA, yet there is no information to indicate which ones are better. Nijkamp (1989) indicates that there could be as many as 50 possible evaluation methods. To date, there has been relatively little research to evaluate the performance of these methods.

Although MCA does not necessarily require quantitative or monetary data, the information requirements to compile the effects table and derive the weights can, nevertheless, be considerable. What is viewed as the method's strength (i.e., use of subjective judgement) may also be viewed as a weakness. Subjective judgements, backed by substantial data sets and information from stakeholders, need to be made in the process and the debatable issue arises as to whose preferences the decision maker represents. Although the weights used in the process are explicit weights, the analyst may unintentionally introduce implicit weights during the evaluation process. If not properly used MCA has the potential to become a 'black box', producing results that cannot be explained.

8.5 Case Study: Application of MCA to Resource Management in Cattle Creek Catchment, Far North Queensland[47]

This section presents an MCA study that was carried out in the Cattle Creek Catchment in Far North Queensland. The outline is as follows. First, a description of the resource problem requiring management is described. Second, a summary outline of the MCA methodology is provided with information and discussion about the involvement of the stakeholders in the decision process. Third, the rankings of the alternatives are presented and discussed. The final section makes some concluding comments about the possibility of a compromise decision being identified from the information provided through the MCA. In addition, some comment is provided about the validity of any decisions from the MCA and about the apparent achievements of the process.

8.5.1 The Problem

Cattle Creek Catchment encompasses approximately 16,700ha of agricultural land in the Mareeba-Dimbulah Irrigation Area (MDIA) and is

---

[47] This case study is taken from *Using a Multiple Criteria Decision Support System to Support Natural Resource Management Decision-Making for Ecologically Sustainable Development*, a PhD thesis by Jackie Robinson, Department of Economics, The University of Queensland, Brisbane, Australia.

situated within the Mitchell River Watershed in Far North Queensland. Land and water resources in the catchment are under increasing pressure. The quality and depth to the groundwater in parts of the catchment is deteriorating whilst at the same time agriculture in the catchment is undergoing restructuring with farmers redeveloping as well as expanding production into crops which promise higher returns. Sustainable development in the catchment, and downstream of the catchment, is at risk unless current and future resource use can be managed to reduce, or at least stabilise, the groundwater problems.

Downstream of the Cattle Creek Catchment and within the MDIA there are approximately 240 irrigators, many growing tobacco which requires irrigation water with less than 25ppm chloride. Some of these irrigators, approximately 31 per cent, would be affected by the quality of the water entering the Walsh River from Cattle Creek. It is estimated that over $2 million of agricultural production per annum could be at risk if the groundwater problems in the Cattle Creek Catchment are not stabilised.

Since the extent of the groundwater problems became apparent, speculation from Downstream Irrigators and from groups of Catchment Irrigators about responsibility for the problem increased. An *ad hoc* approach to addressing the problem has led to conflict between groups of stakeholders including irrigators, both downstream and within the catchment, as well as agency groups. The increased conflict has highlighted the need to adequately inform stakeholders and to encourage their involvement in natural resource management decision-making.

8.5.2   The MCA Approach

The MCA, adapted from a prototype MODSS developed by the U.S. Department of Agriculture, Tucson Arizona (Yakowitz *et al.*, 1992; Shaw *et al.*, 1998), was developed to assist the process of decision-making with respect to natural resource management in the Cattle Creek Catchment.

There are two specific characteristics of the MODSS developed for the Cattle Creek Catchment that make it unique in the development of decision-making processes at a catchment level. First, the groundwater problem in the catchment has been investigated within a multi-disciplinary framework. Second, management alternatives have been identified and stakeholder needs from resource management, including identification of the evaluation criteria and the importance order of the criteria, have been solicited through a survey of stakeholders (Robinson and Rose, 1997).

A land and water data base and a groundwater simulation model of the catchment increased stakeholder acceptance that a resource problem exists in the catchment and, in addition, increased the validity of the proposed alternatives for management and their evaluation.

A survey of stakeholders, in particular those on whom resource management decisions are likely to impact directly, was a valuable part of the decision-making process. The survey provided an opportunity to inform people about the natural resource problems in the catchment and the consequences of inaction and then to solicit their preferences about future management.

Identification of alternatives and criteria

A number of resource management alternatives and evaluation criteria were identified at workshops held in the catchment for stakeholders. Information was solicited from stakeholders, through a survey, about their preferences for the alternatives, as well as to establish a ranking of a number of criteria to evaluate the alternatives.

Scoring the alternatives

It was not expected that individual stakeholders would have a sufficient knowledge base to enable them to score the performance of alternatives against the criteria. As a result, a number of 'experts', considered to be proficient in social, scientific and economic fields, were recruited to score the estimated performance of the alternatives. Alternatives were scored in a variety of measures, including dollars; tonnes per ha; meters per annum; and, a qualitative score was provided to measure the performance of an alternative to reduce degradation downstream of the catchment. The scores were standardised using a technique developed by Yakowitz *et al.* 1992. This technique 'normalised' the scores using a number of generic scoring functions.

Weighting the scores according to the preferences of the stakeholders

Rather than assigning a single set of weights or ranking to the criteria, the criteria were ranked consistent with the preferences of each group of stakeholders. This was important because it was an acknowledgement that each group had a set of preferences that represented their specific needs from resource management.

Evaluating the alternatives

The MCA presents the scores assigned to the management alternatives in an effects matrix showing the performance of each proposed alternative against individual evaluation criteria. Ranking the alternatives according to their scores is valuable information because it demonstrates the trade-offs that are implicit when an alternative is chosen. Table 8.2 shows some of the trade-offs that would need to be considered when making a choice between four of the proposed alternatives. This information is particularly valuable when stakeholders regard themselves to be in conflict with decision-making agencies.

Table 8.2 Trade-offs for the Cattle Creek Catchment resource management problem

	Management Alternatives				
Criteria	Plant trees	Line storage	Efficient irrigation	Restrict water allocations	De-water
Cost to implement	1	5	3	2	4
Downstream impacts	3	4	1	2	5
Increase depth to groundwater	5	4	2	3	1

Source: Robinson (1999a).

Table 8.2 shows that *Plant Trees* performs well against the cost criterion, but its performance in relation to increasing the depth to groundwater is not as impressive. The alternative which performed the best over all the criteria, *Efficient Irrigation*, involves a trade-off in relation to the cost and it is not the best alternative to increase the depth to groundwater. Interestingly, the alternative scored as the most effective in increasing the depth to groundwater, *De-water*, does not perform well against the other evaluation criteria.

Ranking alternatives when the criteria are weighted

A logical progression from inspecting the trade-offs is to rank the alternatives according to the aggregate score, given the ranking of criteria preferred by individual stakeholder groups. The maximum or minimum aggregate score for each alternative, given the ranking of the criteria is given as:

Maximise (minimise) $$\sum_{i-1}^{m} w(i) \cdot Sc(i, j) \qquad (8.3)$$

Subject to: $$\sum_{i-1}^{m} w(i) = 1 \qquad (8.4)$$

and $$w_1 \geq w_2 \geq, ..., \geq w_m \geq 0 \qquad (8.5)$$

where $w(i)$ = weight vector based on the ranking for criterion i, and $Sc(i,j)$ = score of alternative j evaluated or scored for decision criterion i.

Calculation of the maximum and minimum aggregate scores (assuming a given ranking of the criteria) for each alternative enables a ranking of alternatives to be determined. The difference between the maximum and minimum aggregate scores for an alternative also provides information about the sensitivity of the scores to the weightings of the criteria.

Figure 8.4 Ranking of management alternatives for Catchment Irrigators

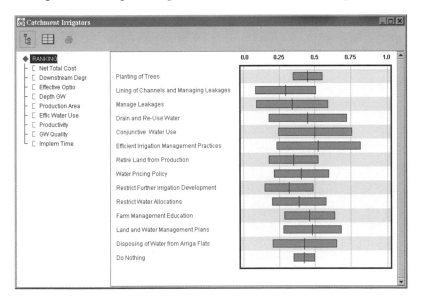

Source: Robinson (1999b).

This information is presented graphically for two groups of stakeholders: Catchment Irrigators (Figure 8.4) and Downstream Irrigators (Figure 8.5). The length of the bars, determined by the ranking of the criteria, shows the difference between the best and worst aggregate scores given the ranking of the criteria by individual stakeholder groups. Although an alternative may perform well in terms of its maximum aggregate score, its performance for the minimum score may be less impressive. The length of the bars is not the same for all alternatives, for all stakeholder groups. For example, *Efficient Irrigation* for Catchment Irrigators (Figure 8.4) and for Downstream Irrigators (Figure 8.5), shows that the ranking of the criteria by each group has a significant effect on the range of aggregate scores and that this range can impact on the ranking of the alternatives.

Figure 8.5 Ranking of management alternatives for Downstream Irrigators

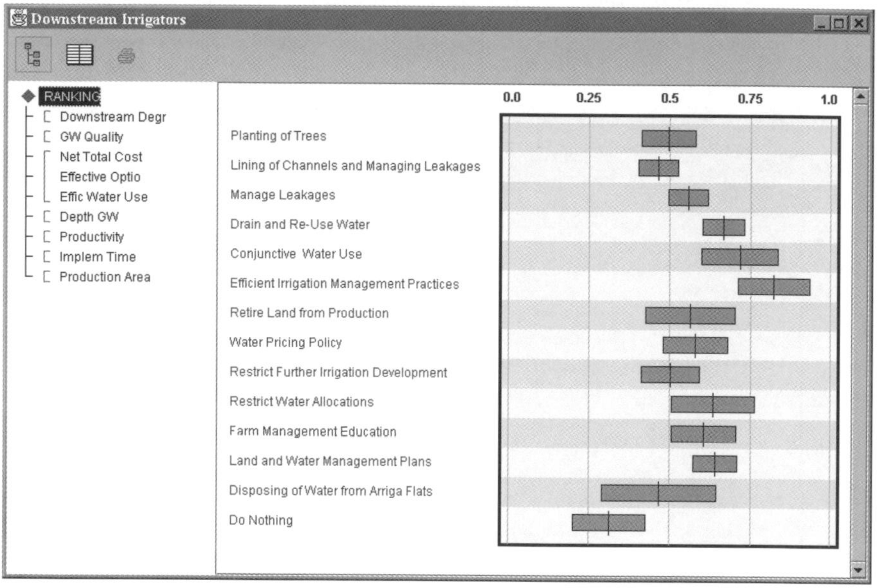

Source: Robinson (1999b).

Figure 8.5 shows the preferred ranking of the criteria for the Catchment Irrigators and the resulting ranking of the alternatives. One end of the graph shows the maximum aggregate score for an alternative and the other end shows the minimum aggregate score. If the maximum aggregate score is used to select the preferred alternative, then *Efficient Irrigation* is ranked first. On the other hand, if the maximum of the minimum scores is used, then the preferred alternative is *Planting Trees*. The ranking of alternatives for the Downstream Irrigators is shown in Figure 8.6 to be different.

A final ranking of the alternatives for each group of stakeholders, shown in Table 8.3, demonstrates that there is a compromise outcome available to stakeholders. In brief, *Efficient Irrigation* is shown to perform the best for all groups of stakeholders using the maximum aggregate score. The final ranking of alternatives shows that *Planting Trees* does not perform well against the evaluation criteria when the criteria are weighted. This is an interesting outcome because it was a preferred alternative for all groups of stakeholders.

Table 8.3 Final ranking of management alternatives, using the maximum aggregate score for each stakeholder group, 1997

Stakeholder Group	Catchment Irrigators	Downstream Irrigators	Community Representatives
Efficient Irrigation	1*	1*	1*
Farm Management Education	4	6	6
Plant Trees	5	11	9
Land & Water Management Plans	3*	4	4
Conjunctive Water Use	2*	2*	2*
Lining Storage	14	13	13
Drain and Reuse Water	5	3*	4
Water Pricing	9	7	7
Retire Land	11	8	7
Restrict Water Allocations	10	4	3*
Manage Leakage	13	9	12
Restrict Irrigation Expansion	12	10	10
De-water	7	12	10
Do Nothing	8	14	14

Note:
* Ranked in the first three.

Source: Robinson (1999a).

8.5.3 Conclusion

The MCA developed for the Cattle Creek Catchment placed considerable emphasis on the need to draw on the preferences of the stakeholders to assist in the determination of an appropriate management alternative or alternatives. Indeed, one of the strengths of the technique is the incorporation of stakeholder preferences within the decision-making process to enable a compromise solution to be identified when there are a number of alternatives to be evaluated and competing and conflicting criteria to be considered. The involvement of stakeholders in the process of decision-making can diffuse a situation or prevent a situation from occurring where there are conflicting opinions and requirements from resource management. Soliciting of stakeholder preferences is used in this study as a means of identifying and overcoming any conflict of interest between stakeholder groups over management of the groundwater resources in the catchment and to gain acceptance of the management alternatives to be implemented (Robinson, 1999a).

The information available to decision-makers from the process, particularly that relating to stakeholder preferences and trade-offs, increases the transparency of the decision-making process. Transparency of the decision increases the validity and acceptability of the choice to all groups of stakeholders.

8.6   Summary

This chapter has introduced the reader to the technique of multi-criteria decision analysis (MCA). This approach is suitable for the analysis of natural resource and environmental problems where there are multiple objectives which may be conflicting or competing. MCA utilises both quantitative and qualitative data and is useful when non-market estimates of environmental effects cannot be determined. Cost-benefit analysis is a special case of MCA when there is a single performance criterion that is measured in monetary terms.

A typical MCA is conducted in 8 steps: identification of the problem; identification of the alternatives; identification of the criteria; scoring of the alternatives; assignment of weights to the criteria; evaluation of the alternatives; sensitivity and risk analyses; and ranking the alternatives. The approach is amenable to input from the analyst, decision maker and stakeholders. In practice, there will be iteration between the various steps after consultations with the interest groups. In spite of its obvious strengths, MCA has a number of weaknesses. There is a large number of evaluation methods and there is little guidance as to which ones are better. The approach relies on subjective judgement and there is the possibility that this may not reflect the society's preferences.

The criteria for judging the success of a MCA, particularly when applied to assist natural resource management, should be related to the critical insights gained through improved communication of the different perspectives of researcher, farmer and decision-maker. The effectiveness of a MCA is in the bringing together of different points of view to bear on a common problem.

Key Terms and Concepts

aggregation techniques
consensus maximisation
concordance methods
electre methods
equipreference curve
graphical technique

lexicographic technique
monte carlo simulation
multi-attribute utility model
multi-criteria analysis
trade-off curve

Review Questions

1. Define multi-criteria analysis.

2. List the steps in MCA.

3. Briefly discuss five approaches for evaluating the alternatives in MCA.

4. Name two advantages and disadvantages of MCA.

Exercises

1. Suppose the government is considering constructing a major freeway in a highly populated city. The alternatives are four locations in the city.

    a. Produce a chart similar to Figure 8.1 showing the possible objectives and the criteria to be considered.

   b.   Suggest units in which the criteria could be quantified.

2.   Suppose the government is considering five solid waste disposal options.

   a.   Produce a chart similar to Figure 8.1 showing the possible objectives and the criteria to be considered.
   b.   Suggest units in which the criteria could be quantified.
   c.   Suggest ways in which the weights could be derived.

References

Department of Natural Resources, DNR (1997). *Location of Cattle Creek Catchment in the Michell River Watershed*. GIS & Cartography Unit, Mareeba.

Department of Natural Resources, DNR (1999). *Facilitator*. Resource Science Centre, Indooroopilly.

Forman, E.H. (1990). Multi Criteria Decision Making and the Analytical Hierarchy Process. In C.A. Bana e Costa (ed.), *Readings in Multiple Criteria Decision Aid*. Springer, Berlin.

Janssen, R. (1992). *Multiobjective Decision Support for Environmental Management*. Kluwer Academic Publishers, Dordrecht.

Janssen, R. and van Herwijnen, M. (1991). Graphical Decision Dupport Applied to Decisions Changing the Use of Agricultural Land. In A. Lewandowski and J. Wallenius (eds.), *Multiple Criteria Decision Support*. Springer-Verlag, Berlin.

Munasinghe, M. (1993). *Environmental Economics and Sustainable Development*. World Bank Environment Paper No. 3., The World Bank, Washington, D.C.

Nijkamp, P. (1989). Multicriteria Analysis: A Decision Support System for Sustainable Environmental Management. In F. Archibugi and P. Nijkamp (eds.), *Economy and Ecology: Towards Sustainable Development*. Kluwer Academic Publishers, Dordrecht.

Pearce, D.W. and Turner, R.K. (1990). *The Use of Benefits Estimates in Environmental Decisionmaking*. Report to the Environment Directorate, OECD, Paris.

Resource Assessment Commission, RAC. (1992). *Multi-Criteria Analysis as a Resource Assessment Tool*. RAC Research Paper No. 6, Australian Government Publishing Service, Canberra.

Robinson, J.J. (1999a). Catchment Management: Facilitating Stakeholder Participation. *Water*, November/December: 35-39.

Robinson, J.J. (1999b). Using a Multiple Criteria Decision Support System to Support Natural Resource Management Decision-Making for Ecologically Sustainable Development. PhD Thesis, Department of Economics, The University of Queensland, Brisbane, Australia.

Robinson, J.J., and Rose, K. (1997). *Stakeholder Preferences for Groundwater Management in the Cattle Creek Catchment*, Unpublished project report, Department of Natural Resources, Mareeba.

Roy, B. (1990). The outranking approach and the foundations of the ELECTRE methods. In C.A. Bana e Costa (ed.), *Readings in Multiple Criteria Decision Aid*. Springer, Berlin.

Saaty, T.L. (1980). *The Analytical Heirachy Process*. McGraw Hill, USA.

Shaw, R., Doherty, J., Brebber, L., Cogle, L., and Lait, R. (1998). The Use of Multiobjective Decision Making for Resolution of Resource Use and Environmental Management Conflicts at a Catchment Scale. In S.A. El-Swaify, and D.S.Yakowitz (eds.), *Multiple Objective Decision Making for Land, Water and Environmental Management*. Proceedings of the First International Conference on Multiple Objective Decision Support Systems (MODSS) for Land, Water, and Environmental Management (1996): Concepts, Approaches, and Applications, CRC Press, USA.

Yakowitz, D.S., Lane, L.J., Stone, J.J., Heilman, P., Reddy, R.K., Imam, B. (1992). Evaluating Land Management Effects on Water Quality Using Multi-objective Analysis within a DSS. *Proceedings of the First International Conference on Groundwater Ecology*, July, 1992, Tampa, Florida. USEPA, AWRA, pp.365-374.

Van Pelt, M.J.F. (1991). Sustainable Development and Project Appraisal in Developing Countries. Paper presented to the 31st European Congress of the Regional Science Association, Lisbon, Portugal, 27-30 August.

Part III. Global Environmental Issues

9. Population Growth, Resource Use and the Environment

Objectives

After studying this chapter you should be in a position to:

❏ review the various positions taken in the debate about the impacts of population and economic growth on the environment;

❏ explain the basic relationships between population growth, resource use and environmental degradation;

❏ explain the relationship between poverty and the environment; and

❏ discuss the implications for economic and social policy

9.1 Introduction

It was established in Chapter 1 that the economy and the environment are closely interrelated. In particular, we saw that the environment has an assimilative capacity that can be considered as finite. It is generally believed that the need to clothe and feed a rapidly growing population is the major cause of environmental degradation. In this and the following chapters, we continue to examine in more detail the relationship between population, the economy and the environment. Although concern for the environment has heightened within the last two decades, the debate over population growth and the environment has raged over the past two centuries. The chapter begins with a review of various positions taken on the relationship between population growth and the environment. This is followed by a description of world and regional population growth trends. Next, we examine the relationship between population growth, resource use and environmental degradation, including the role of poverty. The discussion concludes with a set of policy recommendations.

9.2   Growth Pessimists and Optimists

Historically, the debate about population growth and the environment has been polarised between growth pessimists and optimists. Growth pessimists believe that population growth and the consequent increase in demand for goods and services is the main cause of environmental degradation. On the opposite side, there are the growth optimists who believe that technological advancement can mitigate the adverse environmental impacts of growth. We first take a look at the views of the pessimists.

9.2.1   The Growth Pessimists

In 1798, Thomas Robert Malthus, one of the pioneeers of the economics profession, wrote a book entitled, *An Essay on the Principle of Population,* which subsequently set the tone for the current debate about the sustainability of economic growth. Malthus reasoned that there is a tendency for every species, including the human one, to increase at a geometric rate (e.g., doubling every 50 years), whereas 'under the most favourable circumstances usually to be found', its subsistence (i.e., food production) increases at an arithmetic rate (i.e., at a constant rate of, say, 100 tonnes per annum). Malthus identified two types of population checks: 'preventive' and 'positive checks'. He suggested that the main preventive check is 'moral restraint' or the postponement of marriage with sexual abstinence. Positive checks include wars and natural disasters. According to Malthus, 'when population has increased nearly to the utmost limits of the food, all the preventive and the positive checks will naturally operate with increased force...till the population is sunk below the level of the food; and then the return to comparative plenty will again produce an increase, and, after a certain period, its further progress will again be checked by the same causes' (Malthus, 1872:369-370). The Malthusian Hypothesis is illustrated by Figure 9.1. Malthus's predictions earned the economics profession tags such as the 'profession of doom' and the 'dismal science'.

Ricardo (1817) modified the Malthusian model by including the concept of **diminishing marginal productivity**. That is, as population increased people would be forced to not only cultivate existing land more intensively but also extend production into inferior land.

Figure 9.1 The Malthusian Population Model

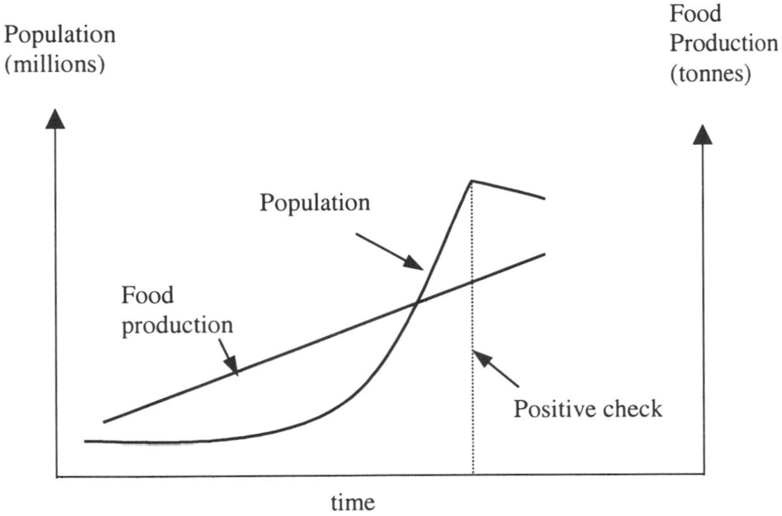

Source: adapted from Goodstein (1995).

Therefore, although total output might increase, it would do so at a declining rate in the absence of technological advancement. The Ricardian view was more optimistic in terms of allowing for mitigation of the Malthusian doom through technical progress.

Malthus's predictions were not taken seriously at the time they were made. The dominant philosophy in the early 19th century was 'mercantilism', which was concerned with amassing wealth for the state. The function of the population was to produce for the state, and to this end, even young children were recruited into the labour force. Although prominent economists such as Jevons and John Stuart Mill expressed concerns about economic growth and resource depletion, natural resources were not generally regarded as a constraint to economic growth. Although natural resources were being rapidly depleted in Britain, the 'Malthusian trap' was 'escaped' by trade in food, as a result of improved transportation which opened up the plains of

North America. After the Second World War, policy makers were preoccupied with reconstruction and with pressing issues such as unemployment and inflation. Acceleration in the growth rate of the economy was viewed as the only solution to these problems.

The advent of modern environmentalism can be traced to the 1960s when the increase in environmental pollution led to a rise in public environmental awareness. After the energy crisis of the 1970s the Malthusian debate was resurrected with neo-Malthusians launching an attack on the virtues of economic growth. In 1972, the neo-Malthusian group, The Club of Rome, published the results of a study entitled, *The Limits to Growth* (Meadows *et al.,* 1972), which attracted extensive media coverage. The study reached three major conclusions. First, at the prevailing annual rates of consumption the world would run out of mineral resources within 100 years resulting in a sudden collapse of the world economic system. Second, this calamity would not be averted by piecemeal solution to the myriad problems. Third, the only solution to the collapse would be an immediate reduction in economic growth, population growth and pollution.[48]

In a report entitled *The Global 2000 Report to the United States President* another neo-Malthusian group argued that increasing population growth and per capita consumption could create severe natural resource shortages by the year 2000 (Barney, 1980). It also predicted that the world would be more susceptible to 'low-probability' and 'high-risk' events such as the 'greenhouse effects' and global nuclear risks. These events would cause widespread disruptions of food production.

The dire predictions of the Club of Rome and the Global 2000 Report led economists including Herman Daly and others to propose a **stationary** or **steady-state economy** in order to avoid prospects of exponential growth. Among other things, Daly recommended a zero population growth and quotas on the use of minerals. He suggested that these quotas could be auctioned to promote economic efficiency.[49] He recommended restricting consumption per capita to a minimum tolerable level per person, arguing that his approach could ensure the longest possible survival of the human species. Paul Ehrlich in his book, *The Population Bomb*, argued that

---

[48] The model follows the basic premise of the Malthusian Theory. Several resources are fixed in supply and growth increases exponentially. These assumptions and the absence of technical progress ensure the exhaustion of resources.

[49] See, for example, Daly (1980) and Erhlich and Harriman (1971).

population is not merely **an** important problem but is **the** problem in ensuring the long-term survival of the human race (Ehrlich, 1970).

In his contribution to the debate, Georgescu-Roegen (1975) argued that zero population growth and a steady-state world economy was not sustainable and that it would eventually lead to a depletion of resources. He based his argument on the Second Law of Thermodynamics, which states that the entropy (i.e., disorder) of a closed system continually increases. According to Georgescu-Roegen, economic activity relies on low-energy entropy. Eventually all energy resources are dispersed and are no longer available for economic production. Although technology could delay the point at which natural resources were completely exhausted, continuing economic activity would ensure their ultimate depletion (Georgescu-Roegen, 1971). He suggested that the only solution to the problem was to reduce the world's population to a level that could be maintained solely on organic agriculture.

The word 'sustainability' was popularised by the International Union for the Conservation of Nature and Natural Resources (IUCN) in the landmark document, *The World Conservation Strategy (WCS)*.[50] The WCS acknowledges the adverse environmental impacts of human kind's economic activities. It draws attention to 'the predicament caused by growing numbers of people demanding scarcer resources [which] is exacerbated by the disproportionately high consumption rates of developed countries' (IUCN, 1980:2).

Although it advocates a precautionary approach, the WCS is, strictly speaking, not growth pessimistic in that it does not advocate any of the extreme measures proposed by the neo-Malthusians (e.g., zero population growth). The document highlights factors which are necessary to achieve sustainable development. These include:

1. Preserving genetic diversity;
2. Maintaining essential ecological processes and life-support systems (eg., agricultural systems, forests, and marine systems), and
3. Ensuring the sustainable use of species and ecosystems.

---

[50] See IUCN (1980).

9.2.2   The Growth Optimists

The *Limits to Growth* hypothesis has been rebutted on several grounds including its lack of consideration for the ways in which technological progress and the market system affect resource scarcity. Friedrich Engels rejected the Malthusian and Ricardian views, arguing that:

> .... science increases at least as much as population. The latter increases in proportion to the size of the previous generation, science advances in proportion to the knowledge bequeathed to it by the previous generation, and thus under the most ordinary conditions also in geometric progression. And what is impossible to science? (Engels, 1959:204).

Simon (1981) took an extremely optimistic view of population growth. According to Simon, population growth is a positive thing because the larger the world's population the more minds there would be and therefore the greater would be the growth of knowledge. This expansion in knowledge would overcome the resource constraints to population growth.

Other optimists who used the power of technology as a basis for their assessments included Kahn *et al.* (1976) and Barnett and Morse (1963). Kahn *et al.* argued that increases in technological progress would enable production and per capita incomes to rise in developing countries and cause their populations to stabilise as has been the case in the developing countries. Barnett and Morse argued that technological progress is automatic, self-reproductive and exhibits increasing returns, and as such, there would be increasing returns to effort.

The release of the Brundtland Commission Report entitled *Our Common Future* (World Commission on Environment and Development, WCED, 1987) was a landmark in the sense that it focussed world attention on sustainable development as a desirable development strategy.[51] The WCED admitted the seriousness of the world's environmental problems but rejected the *Limits to Growth* hypothesis. The Commission argued that the world's resources are sufficient to meet our needs and that a reduction in poverty is one means of reducing environmental degradation. Beckerman (1992) disputed the generally held view that exhaustion of resources and

---

[51] 'Sustainable development' is discussed in Chapter 11.

environmental pollution, in particular greenhouse gas emissions, constrained economic growth. He admitted that economic growth usually leads to environmental degradation in the early stages of development. However, in the end, the best and only way to achieve a decent environment is by wealth accumulation.

9.3 World Population Growth and Trends

The world's population is currently estimated at about 6 billion (Figure 9.2). It is projected to reach nearly 7 billion by 2015 and 8.5 billion by 2040. Population growth rates in the developing countries have accelerated in the last three decades due to factors such as improved health care (e.g., vaccination, antibiotics and anti-malaria measures), sanitation, and education which have decreased mortality rates and increased life expectancy. However, the world's population growth rate is projected to decline from the current 1.6 percent per annum to 1.1 percent per annum by 2040 (Table 9.1).

The figures in Table 9.1 exhibit the so-called **'demographic transition'** hypothesis. That is, lower population growth rates are associated with higher income countries and vice versa. This relationship also holds true for a given country. That is, the poorest families tend to support the highest proportion of children. For example, Birdsall and Griffin (1988) reported that in Columbia, Malaysia and Brazil, the poorest 20 percent of households support about 30 percent of the children in the country.

There are a number of socioeconomic reasons to explain the large family sizes in poor countries. These include the following:

1. In view of the agricultural bias in these countries, there is an incentive to invest in children who are a cheap source of farm labour. Furthermore, the amount of land that can be farmed is a function of family size.
2. Children serve as a form of economic insurance. It is to be noted that most of these countries do not have social security systems. Thus, child labour is an additional source of income for low income families. Investing in children is a way of ensuring care in the parents' old age.

Figure 9.2 World Population, 1980, 1997, and 2015 (millions)

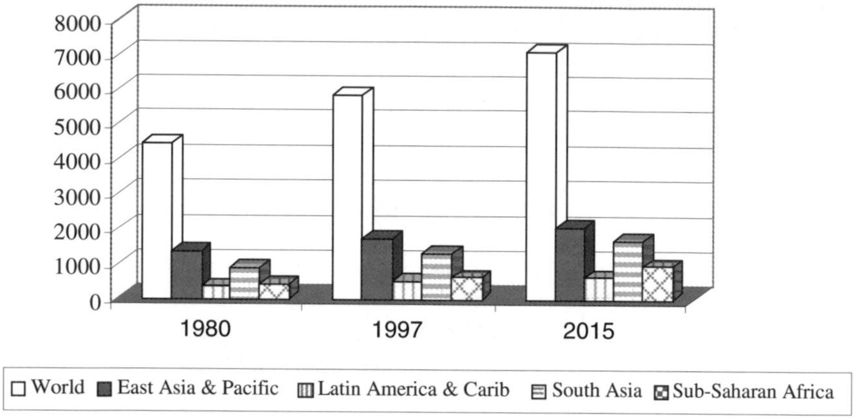

☐ World ■ East Asia & Pacific Ⅲ Latin America & Carib ☰ South Asia ⊠ Sub-Saharan Africa

Source: World Bank (1999).

Table 9.1 World population growth rates, 1980–97 and 1997–2015

Region	Average annual population growth rate (percent)	
	1980–97	1997–2015
World	1.6	1.1
East Asia and the Pacific	1.5	0.9
Latin America and Caribbean	1.9	1.3
South Asia	2.1	1.4
Sub-Saharan Africa	2.8	2.3

Source: World Bank (1999).

3. In many cultures in the developing world, a family's, and in particular a woman's, social standing is directly related to the number of children.

The foregoing suggest that although the factors influencing family size in developing countries are complex, policy should consider the fact that socioeconomic variables play an instrumental role in determining family size. A poor family's decision to have more children is entirely rational in view of high mortality rates and poor economic prospects. Thus, it seems that one way to reduce family size is to improve the economic status of the poor, particularly that of women. This issue is discussed in more detail later.

9.4 Population Growth and the Environment

Population growth increases the demand for goods and services which, in turn, puts additional pressure on environmental resources. The more people there are, the greater is the amount of waste production, and the more adverse are implications for health conditions and the environment's assimilative capacity.

Population growth, poverty and environmental degradation are closely interelated. For example, increasing population leads to more intense use of land, shorter fallow periods and lower soil productivity. It also leads to more clearing of forest cover and hillsides. The net result of these effects is that there is increased environmental degradation (e.g., soil erosion, mudslides, and so on), reduced soil productivity and, hence, lower yields. This results in a fall in per capita incomes and an increase in poverty. The poverty creates a vicious cycle in that it leads to further land degradation as the poor desperately try to eke out a living on the marginal land. However, to borrow the Simon proposition, population growth may have a positive feedback. For example, farmers may be encouraged to adopt technological innovations in agriculture (e.g., the Green Revolution). Population growth increases the labour force, resulting in an increase in the production of goods and services. However, in general, the positive effects are likely to be offset by the negative effects, resulting in a net negative effect. In the next section, the relationship between population growth, poverty and land degradation is examined in a little more detail.

9.4.1   Population Growth and Land Degradation

Land degradation may be defined as a 'loss of land productivity, quantitatively or qualitatively, through various processes such as erosion, wind blowing, waterlogging, depletion of nutrients, deterioration of soil structure and pollution' (Dudal, 1981:4). From this definition, it can be seen that land degradation involves several processes including water and wind erosion, biological degradation (e.g., decrease in humus), physical degradation[52] (e.g., decrease in soil permeability), chemical degradation (e.g., acidification, toxicity) and excess salts (e.g., salination and alkalisation). Soil erosion, the detachment and transportation of soil particles by wind or water, is but one aspect of the overall process of land degradation. Land degradation can occur by natural processes or by human action. To understand the process of land degradation, in general, and soil erosion, in particular, it is necessary to identify the major causal factors. We briefly discuss below the main causes of (non-natural) land degradation, with particular emphasis on soil erosion.

The causes of land degradation can be classified into two broad categories: direct and indirect causes (Figure 9.3). This classification has been simplified for ease of presentation. In reality, there are interactions between all of these factors. For example, deforestation is accelerated by an increase in poverty, population growth or both, and overgrazing can partly be a result of poor land management practices. We briefly discuss each of these factors below.

Direct causes of land degradation

Human activities such as deforestation, inappropriate land management practices and overgrazing can cause land degradation.

Deforestation

Forests protect the land from the impact of rainfall. They enhance the availability of soil nutrients and contribute to soil strength by providing additional cohesion. Deforestation makes the soil more erodible by reducing the organic matter content and the water holding capacity of the soil. There is a reduction of the soil's infiltration rate, leading to increases in run-off and soil erosion.

---

[52] Physical degradation refers to structural breakdown of soil aggregates and reduced soil porosity.

Figure 9.3 Causes of land degradation

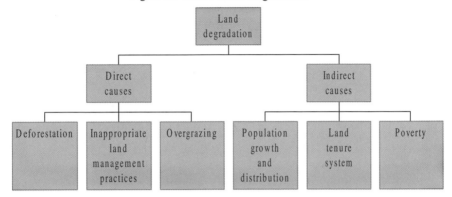

Source: Araya and Asafu-Adjaye (1999).

The massive removal of vegetative cover is a major driving force behind land degradation in many countries today. The rapid increase in rates of deforestation is due to population growth which has increased the demand for cropland, grazing land and wood for construction and fuel.

Inappropriate land management practices

Farming practices such as the types of farm tools used, crops grown, timing of sowing, crop rotation, the use of fertiliser and expansion of agricultural land impact to varying degrees on the rate of soil erosion. Expansion of agricultural land into marginal lands is usually induced by population pressure. In some parts of Ethiopia, Eritrea and Indonesia, population pressure has forced farmers to cultivate steep slopes on hillsides.

Over the past century, technological progress has changed the social, cultural and economic structure of our society and has affected the use of natural resources. Agricultural tools used by traditional societies, although minimally productive and unsophisticated, were consistent with sustainable use of natural resources. The technologies of these societies did not allow large-scale production and exploitation of natural resources. Modern technology has led to an expansion of the scale of economic activities by enabling more rapid clearing of natural habitat and more intensive use of natural resources. However, mechanisation of agriculture has also increased the vulnerability of soils to erosion as traditional cultivation practices such as the use of hoes has resulted in minimum tillage.

It has been suggested that technological progress can have a beneficial effect on conservation of resources. The World Commission for Environment and Development (WCED, 1987) argues that the use of new resource-efficient technologies can increase productivity per unit of a resource, thereby assisting conservation of natural resources. The use of high-yield crop varieties and chemical fertilisers, for example, results in a dramatic yield increase by replacing nutrients that have been depleted by erosion, overgrazing and overcultivation. The increased yield can, in turn, reduce population pressure on land, reduce extension of croplands into marginal lands and thereby reduce the land degradation problem. However, studies carried out in Asian countries indicate that the higher productivity is at the expense of adverse environmental and health effects.[53]

Overgrazing

Crop production in many developing countries is almost entirely dependent on the use of plough oxen. Farmers in some developing countries tend to keep large numbers of livestock as a form of income supplementation and insurance against crop failure. Also, in some countries, ownership of livestock is regarded as a sign of wealth and prestige. Apart from removing vegetative cover, livestock contribute to land degradation by trampling the soil and causing compaction, thereby reducing the rate of infiltration, destroying the aggregate stability of the soil and reducing its water-holding capacity. All of this results in an increase in the risk of soil erosion.

Indirect causes of land degradation

In addition to the factors that directly cause land degradation, other factors also cause or accelerate land degradation indirectly by influencing the agricultural practices of farmers, their decisions about stocking rates, and their willingness and/or ability to make long-term investments in soil conservation. These indirect factors in the causal chain include the following:

- system of land tenure;
- population growth and its distribution; and
- poverty.

---

[53] See studies by Jeyaratnam *et al.* (1987), Antle and Pingali (1994) and Wilson (1999).

System of land tenure

Land tenure has an indirect influence on land degradation in the sense that it influences the type of land use and the adoption of soil conservation measures. A major land tenure problem in many developing countries is the absence, or poor enforcement, of land titles. This insecurity arises from various sources depending on historical patterns of land acquisition and settlement. The absence or the inadequacy of legal and administrative capacity to provide evidence of ownership, and the encroachment of farmers onto common lands, usually contribute to the problem of tenure insecurity.

Insecurity of land tenure caused by poor enforcement of land titles, and the periodic redistribution of farmlands designed to ensure equity among the members of the communities in developing countries, discourages investment in soil conservation. Even when farmers are willing to invest in soil conservation, limited tenure also constrains them from obtaining the loans required to undertake conservation activities.

There is a general belief that communities with communal property systems usually develop a system of resource management that reflect their sense of concern and responsibility for the environment. For example, Pearce and Warford (1993) argue that rural people in developing countries have impressive knowledge of their environment and are able to establish elaborate rules and regulations that enhance sustainable use of their resources. In a study of customary marine tenure (see Appendix 4.1) systems in Papua New Guinea, Asafu-Adjaye (1999) argues that such systems are inherently sustainable as compared to government-imposed fisheries regulations because they are derived from a rich source of environmental knowledge. Nevertheless, a communal management system such as the CMT tends to come under pressure with population growth and technological change. Nadel has made the following remarks with respect to communally owned land in Eritrea:

> "...the spirit of communal responsibility in these communities, makes the temporary landholder work in the interest of his successors as well, because they belong to a closely-knit social unit. The rules of fallow lying and the building and upkeep of terraces which outlive individual tenure, prove this communal spirit convincingly" (Nadel, 1946:14).

It is commonly believed that private ownership of land is more efficient compared to other forms of land tenure in terms of the management of natural resources. The argument is that the security of tenure associated with private ownership of land provides an incentive to farmers to undertake long-term investments such as soil conservation structures and the planting of trees (World Development Report, 1992). However, Pearce and Warford (1993) have argued that private ownership of land may not necessarily lead to conservation of natural resources in developing countries for three reasons. Firstly, the absence of documented land rights does not necessarily imply that land rights do not exist. Many developing countries have historically evolved land and sea rights. Secondly, some forms of private ownership may be associated with unsustainable land use practices. An example is land clearing on agricultural land. Thirdly, title to land is largely meaningless unless it is effectively enforced.

Population growth and distribution

Rapid population growth is often considered to be a cause of land degradation. Most of the population in developing countries depends on the agricultural sector as a means of livelihood. As is most often the case, the other sectors (e.g., manufacturing) offer limited employment opportunities. The extent of the impact of rapid population growth on land degradation depends on the carrying capacity of land and other resources. The land's carrying capacity is defined as 'the maximum population that can be sustained at the minimum standard of living necessary for survival' (Pearce and Warford, 1993:155). As population increases, the per capita area of arable and grazing land decreases, and cultivation extends into marginal lands. The increased demand for cultivable land, firewood and construction materials leads to environmental deterioration. Recent empirical studies suggest that there is a strong correlation between population growth rates and expansion of agricultural land, and between population growth rates and rates of deforestation.[54]

Box 9.1 describes a typical example of the relationship between population growth and deforestation on the island of Java in Indonesia. The population on the island doubled within the past forty years to about 112 million, resulting in massive deforestation. Subsequently, there was large-

---

[54] See studies by Cruz (1994) and Koop and Tole (1999).

scale migration to inland areas of the island, which has resulted in high population densities and further deforestation.

As suggested above, an indirect environmental impact of a high population growth rate is that, as the carrying capacity of the land is approached, it is no longer able to support the population. As a result, there is an increase in rural unemployment and poverty. People then begin to migrate to the towns and cities in search of non-existent jobs. The urban areas themselves do not have the infrastructure to support the rapid influx of migrants. The net result is the development of slums and squatter settlements and the perpetuation of crime, poverty and environmental pollution.

Pearce (1991) has argued that rapid population growth could be a rational response to distinct economic factors such as poverty in rural areas. Thus, although population growth is a major threat to prudent management of natural resources including soil, the mere correlation of population growth with degradation of natural resources does not necessarily imply that population growth is the root cause of the problem.

Population growth itself could be a result of the same factors that cause environmental degradation. The simplified relationship between population growth and land degradation presented above tends to conceal a multitude of factors that contribute to the problem. We briefly examine below the role of poverty in environmental degradation.

Poverty

The relationship between poverty and land degradation is complex. The links between poverty and environmental degradation depend on several factors including the degree of resilience of an area to shocks (e.g., population growth, public policies or climatic changes) and the demands made on natural resources.

Box 9.1 Population growth, migration, deforestation and soil erosion: the Philippines case

The relationship between population growth, deforestation and soil degradation is clearly illustrated by events in the Philippines. The country experienced high rates of population growth for a prolonged period prior to the 1970s, averaging over 3 percent per annum. Although growth has slowed down recently, it was still estimated at 2.4 percent in 1990: one of the highest in the Asian and Pacific Region.

The resulting demand for more arable land for food production led to extensive forest clearing in the upland areas in the country. Estimates indicate that the upland cultivated areas increased almost seven-fold between 1960 and 1987, from 0.3 to 3.9 million ha. Many of the farms were established on slopes as steep as 45 degrees: recent estimates indicate that more than 40 percent of the rural population is located on steep slopes.

With the expansion in cropland acreage, the upland areas attracted an increasing number of landless migrants from the lowlands. This resulted in a net migration to the area, indicated by a population growth rate in excess of the national average. The net upland migration rate more than doubled from 2.1 percent during 1970–75, to over 5.3 percent in 1980–85. At present, the uplands contain more than 20 million people, roughly one-third of the nation's population.

Increased population density in the uplands has resulted in more extensive cultivation, increased cropping frequency and shorter fallow periods. These have led, in turn, to severe soil erosion higher in the catchment areas, resulting in increased siltation and flooding downstream. Soil erosion is estimated at about 122 to 210 tonnes per hectare per annum from newly established farmlands, compared to only 2 tonnes per hectare from forested lands. Preliminary estimates indicate that 90,000 km^2 of the country's upland areas are eroded to a point where they are either totally unsuitable for crop cultivation or can barely support subsistence agriculture.

Source: ESCAP and ADB (1995).

The rural poor in many developing countries depend heavily on their natural resources to cope with a decline in their per capita income caused by any of the above factors. Poverty forces farmers to take a number of actions, including extending cultivation into marginal lands, overgrazing and cutting down trees to sell as firewood. As was explained earlier, such activities often result in soil erosion, deforestation and overgrazing. There is a vicious cycle in the sense that the inability of the degraded land to support the growing population increases the levels of unemployment and poverty.

Empirical studies indicate that the two variables most highly correlated with poverty, in particular in rural areas, are unemployment and lack of (or limited) access to land. For example, in a study of rural areas in India, Sundram and Tendulkar (1993) demonstrated that more than three-quarters of the rural poor and unemployed fell into two classes: landless workers and small-scale farmers. Historically, poverty, defined in absolute as well as relative terms, appears to be more prevalent in rural areas than in urban areas. However, as employment opportunities continue to decrease in the rural areas, migration to the urban areas is increasing the numbers of the urban poor. For example, a study by Prasad and Asafu-Adjaye (1998) indicated that between 1977 and 1990, poverty and income inequality in Fiji increased by 10 percentage points. The analysis indicated that poverty and income inequality increased fastest in urban areas compared to rural areas.

Land is the main resource on which the lives of most farmers in developing countries depend. Even if farmers understand this fact and really care for the land, they face serious constraints in combating land degradation. Poor farmers are also usually marginalised by the fact that they are compelled to cultivate less fertile and steeper slopes with the accompanying high risks of soil erosion. However, these farmers do not have the resources to undertake investments that could enhance the long-term productivity of their land. Poor farmers also cultivate small plots of land and cannot afford to fallow any part of their land.

9.5 Gender Issues

Within the last few decades, there has been increasing recognition of the important contribution that women make in the development process, including the sustainable use and management of natural resources. Many aid-funded programs, as well as national initiatives, now give prominence to the role of women. Many countries' national development programs and strategies now explicitly acknowledge the role of women. Although women comprise about half of the world's population, they are under-represented in the labour force and have higher illiteracy rates compared to men (Table 9.2).

Table 9.2 Gender differences in social indicators

Region	Female population percent of total	Labor force participation ratio of female to male		Adult illiteracy rate, male–female difference		Life expectancy at birth female–male difference
	1997	1970	1997	1970	1997	1997
World	49.6	0.6	0.7	23	15	4
East Asia and Pacific	48.9	0.7	0.8	28	14	3
Europe and Central Asia	51.9	0.9	0.9	11	4	9
Latin America and Caribbean	50.4	0.3	0.5	7	2	6
South Asia	48.5	0.5	0.5	28	27	1
Sub-Saharan Africa	50.5	0.7	0.7	19	16	3

Source: World Bank (1999).

These differences are larger in developing countries than in the developed countries. In addition, women in developing countries are likely to have lower health and nutritional status. Due to their low wealth and employment status, many women do not have access to credit and other support services.

It was stated earlier that countries tend to have lower population growth rates as per capita incomes rise. A case in point is South Korea where the population growth rate fell from 2.4 percent between 1960–70 to 0.9 percent in 1990. However, an important factor in the reduction in population growth is that women entered the workforce at a rapid rate (Repetto, 1985). The economic explanation for this change is that increasing education and employment opportunities for women increase the opportunity cost of having children and reduce the birth rate, in the sense that women delay marriage or reduce the number of desired children, if already married. Another factor is that as families become more educated their incomes increase, and the opportunity cost of the parents' time increases, with options such as work or leisure becoming available. There is a tendency for such families to go for a 'quality' strategy in terms of family size. This means having fewer children and investing more in their education and upbringing.

9.6   Policy Responses to the Population Problem

Many countries have adopted national population policies or strategies that aim to stabilise or reduce their populations. Notable examples are China, India and Indonesia. China, for example, is noted for its one-child policy for urban families. The policy is enforced with a combination of subsidies, fines and, on occasions, physical coercion. The current data suggest that the Chinese strategy has been effective in reducing the growth rate of the population.

In the 1970s the Indian government introduced an aggressive birth-control strategy which provided financial incentives for sterilization. The program was discontinued as a result of public opposition and outcry. It is debatable whether such coercive policies will work in many developing countries. Policies that address the issues of poverty reduction, public health, education and improvement in the status of women are likely to be more successful. We briefly discuss some of these strategies below.

9.6.1   Public Health

Improving public health actually increases population growth rates in the short-run, as morbidity and mortality decline. However, reduced child and infant mortality will, in the long run, encourage families to invest in 'quality' rather than 'quantity', as far as family size is concerned. Over the last three decades, much progress has been made in developing countries in terms of provision of basic sanitation and health services. These developments have resulted in reductions in infant and child mortality and increased life expectancy. However, there still remain major differences between various countries. Within a given country, there are inequalities in access to basic health care between urban and rural dwellers, and between men and women.

9.6.2   Education

Education is highly correlated with income. As families become more educated they tend to earn more, and the opportunity cost of taking time off to raise children increases. Therefore, once again, such families tend to opt for 'quality' in terms of having fewer children and investing more in their education. The current statistics suggest that the provision of basic education in many developing countries is lacking due mainly to shrinking government budgets for social services.

9.6.3   Status of Women

As already stated above, women are often disadvantaged in a male-dominated world. One of the ways in which the status of women can be improved is by implementing policies that improve their educational status and enhance their entry into the workforce. A good example of an initiative that empowers women, by giving them access to credit, is the Grameen Bank in Bangladesh. As stated earlier, many developing countries acknowledge the role of women in national policy statements. Some have set up special ministries to address the needs of women and implement special programs for them. There is the need to ensure that women are adequately represented in the national decision-making process. Although women constitute about half the world's population, in many countries (including advanced countries), they are underrepresented in institutions such as the legislature, the judiciary, public administration and the private sector. Realistically, there is a limit to what the government or aid agencies can do to enhance the status of women. Ultimately, improving the status of women will be faster if the economy is growing. The current economic slow-down in some developing countries implies that efforts to enhance the status of women will also be retarded.

9.6.4   Poverty and Income Inequality

The situation with poverty is similar to that of the status of women. Governments and aid agencies can implement poverty alleviation programs (and many such programs have been implemented or are in the process of implementation). However, where poverty is widespread, there is a limit to what such programs can achieve. It may be said that an effective way to reduce poverty is to have a healthy and growing economy. For example, poverty has fallen drastically in the newly industrialised economies (NIEs) such as Hong Kong, South Korea and Taiwan. Emerging economies such as Thailand and Indonesia have also experienced impressive declines in the numbers of absolute poor.[55]

The issue of income distribution is a different matter because past inequities can become entrenched even when the economy is growing. Some countries (e.g., Zimbabwe) have attempted to redistribute wealth by taking land away from wealthy farmers and redistributing it amongst poor, landless farmers. This form of redistribution has been found to be beneficial in terms of stemming the rural-urban population drift (Moene, 1992). Forster (1992) has proposed an interesting redistribution of wealth between rich and poor countries. Calling it 'debt-for-farmland swaps', she proposes that poor countries be granted debt relief in exchange for funds for land reform. These funds could then be used to finance resettlement of farmers from environmentally fragile areas to more productive land. However, the feasibility of such a scheme has not been demonstrated.

9.7 Summary

Whether economic growth will increase or reduce resource scarcity is not a clear-cut issue. Some of the neo-Malthusian predictions about resource availability have not materialised to date. This is mainly due to the fact that they confined their definition of scarcity to physical terms. Related factors such as the role of the price mechanism, increased resource exploration, improved production techniques, the use of substitutes, research into and development of substitutes, as well as possible changes in consumption patterns were not considered.

On the other hand, some of the views of the optimists are also unrealistic. For example, no level of technological advancement can compensate for some of the handiwork of nature. It is highly unlikely that technical knowledge will reach the stage where pristine natural environments and genetic diversity can be totally recreated. In view of the uncertainty surrounding likely future environmental impacts and the possible irreversible damage to the environment, a pragmatic strategy to follow would be one

---

[55] The Asian financial crisis of 1997 may have temporarily increased poverty to some degree. However, we show in Chapter 10 that most of these countries have emerged from the recession.

based on the concept of sustainable development, which is discussed in Chapter 11.

Population growth can have both a positive and negative effect on environment and development. On the one hand, population growth can force adaptive and technological change and thus contribute to development. On the other hand, population growth contributes to the depletion of natural resources, including reducing development and environmental quality. It was also argued that widespread use of chemical fertilizers and pesticides can cause adverse environmental and health effects. Therefore, on balance, the overall effect of population growth tends towards the negative. Policies which address the underlying economic and cultural factors affecting large family sizes have relatively better chances of success. These policies include improving public health, education, the status of women and addressing poverty and income inequality issues. In the final analysis, an effective way to control population growth is through rapid economic growth and development. While this may sound contradictory, the argument is that economic growth creates employment and leisure opportunities which, in turn, raise the opportunity costs of having more children and results in smaller family sizes.

To conclude, the point must be made that it is not only population growth that contributes to environmental degradation. Sometimes misguided government policies such as under-pricing of timber, subsidisation of land clearing, and agricultural subsidies indirectly contribute to environmental degradation.

Key Terms and Concepts

demographic transition
diminishing marginal productivity
green revolution
growth optimist
growth pessimist
land degradation
land tenure

Malthusian trap
stationary or steady-state economy
biological degradation
physical degradation
chemical degradation
soil erosion

Review Questions

1. Explain the Malthusian population model.

2. Explain how Ricardo's model differs from Malthus' model.

3. What were the main conclusions of the *Limits to Growth* Report?

4. Use the Second Law of Thermodynamics to explain why even zero population growth and a steady-state world economy may not be sustainable.

5. Give three reasons that explain why population growth has accelerated in developing countries over the last three decades.

6. Give three reasons that explain why poor families are likely to have large family sizes.

7. Make a list of the positive and negative environmental impacts of population growth.

8. Explain four causes of land degradation.

9. List and explain issues that need to be addressed in order to reduce population growth.

Exercises

1. To date Malthus's predictions have not come to pass. Why do you think that is so?

2. 'The problem of population is not a question of numbers but a question of inequity in distribution of resources.' Do you agree or disagree? Discuss.

3. Is family planning the best way to control population? Discuss.

4. Give reasons that explain why coercive policies may not be an ideal way to control population growth.

5. Why is improving public health one of the ways to address the population problem when it actually increases population growth in the short run?

6. List the possible advantages and disadvantages of a 'debt-for-farmland swap'.

References

Antle, J.M. and Pingali, P. (1994). Pesticides, Productivity, and Farmer Health: A Philippine Case Study. *American Journal of Agricultural Economics*, 76: 418-430.

Asafu-Adjaye, J. (1999). Customary Marine Tenure Systems and Sustainable Fisheries Management in Papua New Guinea. *International Journal of Social Economics*, 27: 917-926.

Araya, B. and Asafu-Adjaye, J. (1999). Returns to Farm-Level Soil Conservation on Tropical Steep Slopes: The Case of the Eritrean Highlands. *Journal of Agricultural Economics*, 38(2):164-175.

Barnett, H.J. and Morse, C. (1963). *Scarcity and Growth*. Johns Hopkins University Press, Baltimore, MD.

Barney, G.O. (1980). *The Global 2000 Report to the President of the U.S. Entering the 21st Century*. Pergamon Press, New York.

Beckerman, W. (1980). Economic Growth and the Environment: Whose Growth? *World Development*, 20(4): 481-496.

Birdsall, N.M. and Griffin, C.C. (1988) Fertility and Poverty in Developing Countries. *Journal of Policy Modelling*, 10(1):29-55.

Cruz, M.C.J. (1994). Population Pressure and Land Degradation in Developing Countries. In *Population, Environment and Development*. United Nations, New York.

Daly, H.E. (1980). *Economics, Ecology, Ethics*. Freeman, San Fransisco.

Dudal, R. (1981). An Evaluation of Conservation Needs. In R.P.C. Morgan (ed.), *Soil Conservation Problems and Prospects*. John Wiley, Chichester.

Economic and Social Commission for Asia and the Pacific (ESCAP) and Asia Development Bank (ADB) (1995). *1995 State of the Environment in Asia and the Pacific*. United Nations, New York.

Engels, F. (1959). Outline of a Critique of Political Economy. In Karl Max (ed.), *Economic and Philosophical Manuscripts of 1844*. Foreign Languages Publishing House, Moscow.

Ehrlich, P.R. (1970). *The Population Bomb*. Ballantine, New York.

Ehrlich, P.R. and Harriman, L. (1971). *How to be a Survivor*. Ballantine, New York.

Forster, N.R. (1992). Protecting Fragile Lands: New Reasons to Tackle Old Problems. *World Development*, 20(4): 371-382.

Grossman, G.M. and Krueger, A.B. (1991). Environmental Impacts of a North American Free Trade Agreement. *National Bureau of Economic Research Working Paper Series*, No. 3914.

Georgescu-Roegen, N. (1971). *The Entropy Law and the Economic Process*. Harvard University Press, Cambridge, Mass.

International Union for the Conservation of Nature and Natural Resources, IUCN (1980). *The World Conservation Strategy: Living Resource Conservation for Sustainable Development*. Glands, Switzerland.

Jeyaratnam, J., Luw, K.C., and Phoon, W.O. (1987). Survey of Acute Pesticide Poisoning among Health Workers in Four Asian Countries. *Bulletin of the World Health Organisation*, 65(4): 521-527.

Kahn, H., Brown, W., and Martel, L. (1976). *The Next 200 Years: A Scenario for America and the World*. William Morrow, New York.

Koop, G. and Tole, L. (1999). Is there an Environmental Kuznets Curve for Deforestation? *Journal of Development Economics*, 58: 231-244.

Malthus, T.R. (1872). *An Essay on the Principle of Population*. 7th Edition, Reeves and Turner, London.

Meadows, D.H., Meadows, D.L. and Behrens, W. (1972). *The Limits of Growth: A Report to the Club of Rome's Project on the Predicament of Mankind.* Universe Books, New York.

Moene, K.O. (1992). Poverty and Land Ownership. *American Economic Review,* 82(1):52-64.

Nadel, S.F. (1946) Land Tenure on the Eritrean Plateau. *Africa: Journal of the International African Institute,* 16(1).

Pearce, D. (1991). *Blue Print 2: Greening The World Economy.* Earthscan Publications, London.

Prasad, B.C. and Asafu-Adjaye, J. (1998). Macroeconomic Policy and Poverty in Fiji. *Pacific Economic Bulletin,* 13(1): 47-56.

Repetto, R. (1985). Population, Resource Pressures, and Poverty. In Robert Repetto (ed.), *The Global Possible: Resources, Development and the New Century.* Yale University Press, New Haven:

Ricardo, D. (1817). *The Principles of Political Economy and Taxation.* Reprint 1955, Dent, London.

Simon, J.L. (1981). *The Ultimate Resource.* Princeton University Press, Princeton.

Sundaram, K. and Tendulka, S.D. (1993). *Towards an Explanation of Regional Differences in Poverty and Unemployment.* Delhi School of Economics, Working Paper No. 237, University of Delhi, Delhi.

Wilson, C. (1999). *Environmental and Human Costs of Commercial Agricultural Production in South Asia.* Department of Economics, Economic Issues Paper No. 8, The University of Queensland, Brisbane.

World Bank (1992). *World Development Report 1992: Development and the Environment.* Oxford University Press, New York.

World Bank (1999). *World Development Indicators 1999.* CD-ROM Version. World Bank, Washington D.C.

World Commission on Environment and Development (1987). *Our Common Future.* Oxford University Press, Oxford.

10. Economic Growth and the Environment

Objectives

After studying this chapter you should be in a position to:

❑ describe trends in world trade and environmental indicators;

❑ describe the relationship between energy consumption and social welfare;

❑ explain the interrelationships between trade and the environment;

❑ explain the relationship between economic growth and biodiversity; and

❑ propose policy solutions for the adverse environmental impacts of trade.

10.1 Introduction

World trade has been expanding at a rapid pace for several decades, even outpacing global production. In the report, *Our Common Future*, WCED noted that:

> "The planet is passing through a period of dramatic growth and fundamental change. Our human world of 5 billion must make room in a finite environment for another human world. The population could stabilize at between 8 billion and 14 billion sometime in the next century. According to UN projections economic activity has multiplied to create a

US$413 trillion world economy, and this could grow five or
tenfold in the coming century" (WCED, 1987:4).

The adverse impact of economic activity on the environment has been a
subject of much debate and controversy since the 1700s. Aspects of this
debate were briefly reviewed in Chapter 9. Environmentalists argue that
international regulation of trade is necessary to 'build environmental
responsibility into economic activity' and to ensure that 'trade meets the
goals of environmentally sustainable development' (Hair, 1993). As trade
has become globalised, environmentalists claim that the magnitude of
environmental degradation has worsened. According to Herman Daly,
'further growth beyond the present scale is overwhelmingly likely to
increase costs more rapidly than it increases benefits, thus ushering in a new
era of 'uneconomic growth' that impoverishes rather than enriches' (Daly
and Cobb, 1989).

In general, environmentalists oppose free trade from the viewpoint of
'market failure'.[56] According to this view, because the market 'fails' to
protect environmental values to the desired degree, there is a need for
government intervention. Many environmentalists view economic growth as
being incompatible with the maintenance of environmental quality and
therefore advocate political constraints on economic activity both
domestically and internationally. It is true that all economic activities do
impact on environmental quality to some degree. However, government
intervention in every case would be impractical and may not be optimal.

On the other side of the debate, there are those who believe that
economic growth is necessary in order to achieve a cleaner environment and
eradicate poverty. Numerous empirical studies have been conducted to prove
that there is a positive link between economic growth and environmental
quality. In this chapter, we critically review the relationship between
economic growth (or trade, a major component of growth) and the
environment. In light of the evidence presented, we address what policy
measures could be taken to mitigate the environmental impacts of economic
growth.

---

[56] Herman Daly suggests that a more accurate name for 'free trade' is 'deregulated
international commerce' (Daly, 1993).

10.2 Economic and Environmental Trends

We set the discussion into context by considering recent trends in world trade and environmental indicators. We first look at economic trends within the last three and a half decade and then consider environmental trends.

10.2.1 Economic Trends

The statistics displayed in Table 10.1 indicate that the world economy grew steadily between 1965 and 2001 at an average rate of 3.5 percent per annum. However, there was regional imbalance in the distribution of this growth. In particular, the East Asian (and Pacific) region economies grew at an average rate of 7.2 percent per annum. In contrast, the Sub-Saharan African (SSA) countries grew at a rate of 3 percent per annum on average, while Latin American and Caribbean countries grew at 3.9 percent per annum on average.

Table 10.1 Economic trends, 1965–2001

	Average GDP growth p.a. (%)		Average value added growth p.a. (%) (from 1972)		
	Total	Per capita	Agriculture	Industry	Services
World	3.5	1.7	2.0	2.6	3.5
East Asia and Pacific	7.2	5.3	3.7	9.5	7.2
Latin America and Caribbean	3.9	1.7	2.8	3.7	4.2
South Asia	4.5	2.3	2.7	5.3	5.6
Sub-Saharan Africa	3.0	0.3	2.2	3.0	3.4

Source: World Bank (2003).

These figures are brought into sharper contrast when we look at them in per capita terms. For example, the East Asia and Pacific region outperformed the rest of the world with average per capita GDP growth of 5.3 percent per annum, while SSA countries grew at an average of 0.3 percent per annum per capita. For all the regions shown, the main engines of growth were the industrial and services sectors.

The Asian crisis financial crisis of 1997 and natural disasters, including the El Niño[57], caused East Asian and Pacific region economic growth to drop sharply from 9.6 percent in 1991–96 to 4.3 percent in 1997–99 (see Table 10.2). The only exception was the Chinese economy which continued to recorded robust growth of 7 percent per annum during this period. The Asian financial crisis had a ripple effect on the rest of the developing world. East Asia had been responsible for much of the recent growth in global consumption of commodities, with Japan accounting for about a third of world consumption. A drastic fall in Japanese demand, coupled with the lagged supply response to the commodity boom of 1994–96, forced commodity prices to trend downwards. In view of the fact that many developing countries depend on primary commodities for most of their export earnings, growth in these regions slowed considerably. The statistics suggest that the East Asian economies haved emerged from the crisis, with average growth of 6.3 percent per annum in the period 2000 to 2001.

Table 10.2 World real annual gross domestic product growth, 1965–2001

	1965–90	1991–96	1997–99	2000–01
World	3.9	2.4	2.8	2.5
East Asia[a]	7.0	9.6	4.3	6.3
Latin America and the Caribbean	4.1	4.5	2.2	2.2
South Asia	4.2	5.4	5.4	4.6
Sub-Saharan Africa	3.3	1.8	2.7	3.2

Note: [a] Includes Indonesia, Republic of Korea, Malaysia, the Philippines and Thailand.

Source: World Bank (2003).

10.2.2 Environmental Trends

As discussed in the previous chapter, a growing population requires an increase in the supply or production of materials for housing, clothing and feeding. In most cases, the use or extraction of these materials either results in the depletion of environmental resources or a reduction in environmental

---

[57] El Niño is a term used to describe unusually warm ocean conditions along the west coast of South America which adversely affect fisheries and agriculture in the Asia-Pacific region and parts of North America.

quality. As already mentioned, the increase in the rate of depletion of natural resources in the rural areas accelerates the rate of rural-urban migration, which results in other problems such as pollution, unemployment and crime in the urban areas. At present, the rate of rural-urban migration in developing countries is quite high. For example, it is projected that by 2015, there will be 21 megacities[58] in the Asia-Pacific region (ESCAP and ADB, 1995). These changes will put immense pressure on the environment in terms of the provision of energy, water, materials, and food.

Figure 10.1 presents the trends in land use (arable land in hectares per person) for India, Indonesia, the Philippines and China for the period 1980–2000. It can be seen that land use, defined in terms of the amount of arable land in hectares per person, has declined over the period under review, with the fastest rate of decline occurring in India. Arable land in India declined by nearly 50 percent in per capita terms (a fall of 2.3 percent per annum) over this period. With increase in population in all of these countries, it can be inferred that a limited land area is being used to support a growing population. Of course, such a situation could be sustained with improved technology. However, the rate of technological progress in these countries is quite low in relation to the advanced countries.

Figure 10.2 presents trends in the amount of water pollution generated from the pulp and paper industries in the selected countries for 1980–2000. The rate of water pollution from these industries is directly related to the rate of deforestation in the respective countries. It can be seen that water pollution from pulp and paper production in India and Malaysia has stabilized and even fallen below 1980 levels. However, water pollution pollution from these sources in Indonesia and China has increased steadily since 1990, although there has been a decline since 1997.

10.3 Energy Consumption, Economic Growth and Welfare

To the extent that energy is a major input in industrial production, it is to be expected that there will be a strong relationship between energy use and industrial output or economic growth. Figure 10.3 presents a plot of GDP per capita (adjusted for purchasing power parity, PPP, in 1995 US dollars) and

---

[58] A 'megacity' is defined as a city with a population of over 10 million.

Figure 10.1 Land use for selected Asian countries, 1980–2000

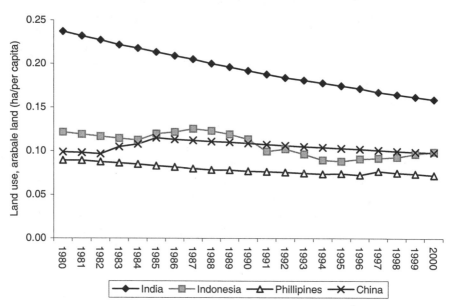

Source: World Bank (2003).

Figure 10.2 Water pollution for selected Asian countries, 1980–2000

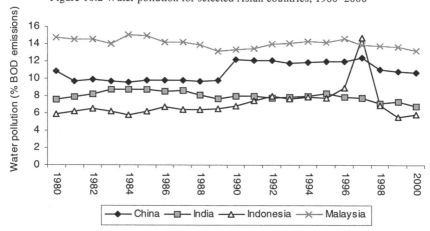

Source: estimated with data from World Bank (2003).

Figure 10.3 Gross domestic product and energy use in selected
Asian countries

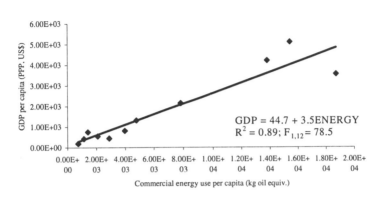

$$GDP = 44.7 + 3.5ENERGY$$
$$R^2 = 0.89; F_{1,12} = 78.5$$

Source: estimated with data from World Bank (2003).

commercial energy use per capita (in kilograms of oil equivalent) for a
sample of 12 countries.[59] The plot indicates a linear relationship between the
two variables. The regression statistics (R^2 and F-statistics) suggest that the
relationship is highly significant. These results suggest that rapid economic
growth in the Asia-Pacific region will entail a rapid development in energy
systems. The converse of this relationship is that curtailment of energy use
will severely retard economic growth.

Energy use is also related to various indicators of social welfare. Figure
10.4 indicates that life expectancy (represented by the variable LIFE) is
linearly related to energy use (represented by the variable ENERGY).
Although the relationship is not as strong as that of GDP versus energy use,
it is nevertheless significant. Figure 10.5 indicates an inverse relationship
between infant mortality and energy use. In both cases, the relationship can
be explained by the fact that energy enhances the development process,

---

[59] The countries were: Thailand, Japan, Republic of Korea, New Zealand, Australia,
Bangladesh, China, India, Indonesia, Malaysia, the Philippines and Sri Lanka. The data were
obtained from World Bank (1999).

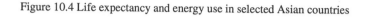

Figure 10.4 Life expectancy and energy use in selected Asian countries

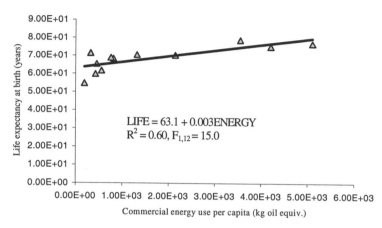

Source: estimated with data from World Bank (2003).

Figure 10.5 Infant mortality and energy use in selected Asian countries

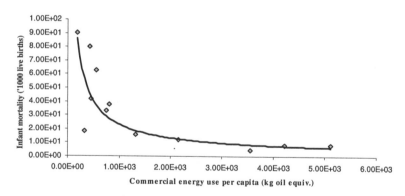

Source: estimated with data from World Bank (2003).

enabling a country's standard of living to be improved. To achieve sustainable development, there is the need to search for efficiencies in energy use and to gradually mover towards renewable forms of energy such as wind, thermal and solar energy.

10.4   The Benefits of Trade: Theoretical Arguments

The economic argument for free trade is based on the theory of **comparative advantage** which says that both partners in trade gain by specializing in goods and services they can produce most efficiently. However, this basic theory ignores environmental externalities associated with trade liberalisation. Take the example of the importation of a good, say, coal (Figure 10.6). Let S be the marginal private cost of producing coal and S' the marginal social cost. Note that the marginal social cost includes both private costs and externalities. In the absence of trade, P^* will be the equilibrium price and Q^* units of coal will be supplied. With free trade, the price of coal will be P_w, which will also be the domestic price; Q_1 units will be produced domestically, while $(Q_1 - Q_2)$ units will be imported. The loss in producer surplus will be equivalent to area P^*bcP_w because domestic producers now sell less coal at a lower price. However, the gain in consumer surplus will be area P^*bgP_w because domestic consumers can now purchase more coal at the same lower price. The net benefits from trade are therefore given by area bcg.

The above analysis does not consider the environmental externalities associated with trade. If coal production damages the environment, the country gains area by $ahbc$ by reducing production from Q^* to Q_1 units. This reduced environmental cost is now shifted onto other countries producing coal for export.[60] On the other hand, if local consumption of coal causes environmental damage, the true marginal benefits from consumption will fall to D', increasing the negative externalities by area $befg$.

The foregoing analysis can be repeated for a commodity, say, timber, that is produced for export (Figure 10.7). In the absence of externalities, timber production increases from Q^* to Q_2 units, while domestic timber consumption decreases from Q^* to Q_1 units, leaving $(Q_2 - Q_1)$ units for export. Total output of timber is therefore Q_2 units. Producers gain area P_wbcP^* by selling more timber at the higher world price P_w. However, consumers lose area P_wacP^* because they can now buy less timber at the higher world price. Therefore, the net gains from trade in the absence of externalities is area abc.

---

[60] This situation is referred to as exporting pollution.

Figure 10.6 Gains and losses from imports

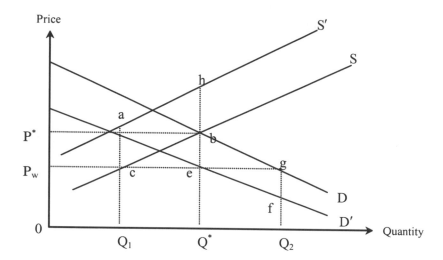

The externalities associated with additional timber production include increased soil erosion and its associated costs (e.g. loss in crop productivity, increased chances of flooding) and loss of ecosystem services such as climate control, water regulation, nutrient recycling and provision of genetic resources. These losses are given by area *debc*. The net gain (or loss) from trade in the presence of externalities depends on whether or not area *abc* is greater than area *debc*. Therefore, in this case, we cannot unambiguously conclude that timber exports are beneficial to the country.

It must be noted that the above free trade model is a simplistic one. It is cast in a partial equilibrium framework and does consider the effect of a policy on the position of the private supply, social cost and private demand curves. It also does not consider interaction between product and factor markets or the preferences of consumers and producers. The empirical evidence on the environmental effects of increased trade is mixed. Using a general equilibrium model, Chichilnisky (1994) concluded that trade increases overexploitation of resources and that establishing private property rights to the environmental resource offers the only feasible response to the problem of overexploitation.

Figure 10.7 Gains and losses from exports

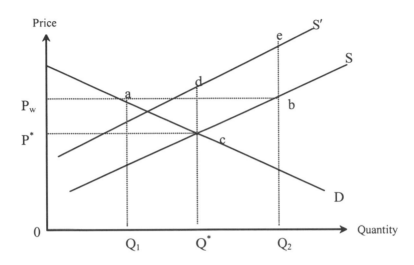

However, Beghin *et al.* (1997) found no wholesale environmental degradation in Mexican agriculture as a result of free trade, while Bandara and Coxhead (1999) found that there were some environmental gains from trade liberalisation in the case of Sri Lanka.

Lutz (1992) projects increased environmental pressures on land, water and soils in developing countries following trade liberalisation. Trade liberalisation could also result in indirect negative environmental impacts such as when small-scale or subsistence farmers are displaced by larger scale export agriculture onto marginal lands. However, Anderson (1992) argues that an expanding commercial labour force in agriculture could help reduce migration to urban areas and therefore reduce urban environmental problems. Also, the higher incomes would reduce the incentive for subsistence farmers to cultivate marginal land or to deforest them for fuelwood.

In general, trade liberalisation increases transport related environmental externalities. For instance, transport of bulky agricultural products is energy intensive and therefore costly in both environmental and financial terms. On the other hand, trade liberalisation could deliver some positive environmental impacts in the form of transfer of environmentally friendly technology to developing countries. It could also promote more efficient production that could reduce material and energy use per unit of output. Trade liberalisation could also compel a country to improve environmental standards when product quality or transboundary effects are at issue. This has already occurred to some extent in Fiji where producers responded to the banning of the fumigant ethylene dibromide, a carcinogen, by switching to non-chemical quarantine treatment technology. Already, some Fijian farmers are beginning to take advantage of significant and growing markets for certified organic products.

10.5   The Relationship Between Economic Growth and Pollution

The term 'Environmental Kuznets Curve' (EKC) was first used by Selden and Song (1994) when they suggested that the environment-income relationship might be similar to the one proposed by Simon Kuznets, the Nobel laureate, for income inequality in relation to development, namely an 'inverted-U' shape (Kuznets, 1955). The basic premise of the EKC hypothesis is that there is an 'inverted-U' shaped relationship between a variety of indicators of environmental degradation and the level of income per capita. That is, as per capita income increases, environmental degradation will initially increase, but then eventually decline once a maximum level is reached.

An example of an EKC for sulphur dioxide (SO_2) is presented in Figure 10.8. A typical feature of the EKC is the inverted U shape that suggests that pollution reaches a maximum with respect to income, after which it begins to decline. The maximum level of pollution is referred to as the **turning point** and forms the focus of the debate about pollution control. If in fact the EKC hypothesis is true, then this turning point will set a per income capita benchmark for developing countries. That is, it is to be expected that developing economies will increase their levels of pollution and environmental degradation until the turning point is reached. The EKC concept can be traced to Vernon Ruttan who remarked in his 1971 presidential address to the American Agricultural Economics Association that:

"…in relatively high-income countries the income elasticity of demand for commodities and services related to sustenance is low and declines as income continues to rise, while the income elasticity of demand for more effective disposal of residuals and for environmental amenities is high and continues to rise. This is in sharp contrast to the situation in poor countries where the income elasticity is high for sustenance and low for environmental amenities" (Ruttan, 1971:707–708).

Figure 10.8 A hypothetical EKC for sulphur dioxide

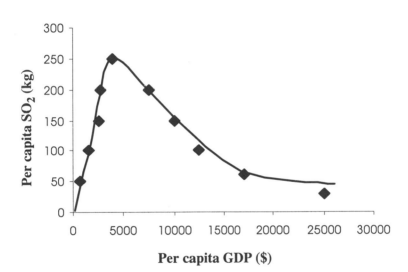

Source: Asafu-Adjaye (1999).

In this section, we first review theories that attempt to explain the EKC relationship. This is followed by a review of empirical studies of the EKC. Next, we conduct a critical assessment of the evidence.

10.5.1 Theories Explaining the Environmental Kuznets Curve

A number of theories have been proposed to explain the inverted U-relationship between economic growth and environmental quality. Here, we take a brief look at three groups of theories: (i) overlapping-generations models; (ii) production/consumption models of pollution, and (iii) political economy models.

Overlapping-generations models

John and Pecchenino (1994) use Samuelson's (1958) and Diamond's (1965) **overlapping-generations**[61] framework to explain why an inverted U relationship might exist between economic growth and environmental quality. In this approach, short-lived individuals make decisions about the accumulation of capital and the provision of a public good, environmental quality, where the decisions have long lasting effects. In the stylized model, economic agents live two periods, working young and consuming while old. The young allocate their wages between investment in capital goods and investment in the environment, which is a public good. Economic agents derive utility from consumption and environmental quality. Their consumption degrades the environment which is left to future generations. However, investment in capital improves the technology available to future generations.

John and Pecchenino (1994) indicate that economic agents in economies with little capital (or high environmental quality) may choose not to maintain the environment. As agents accumulate capital, the consumption externality causes degradation of the environment, resulting in a negative correlation between economic growth and environmental quality. On the other hand, in economies with high capital levels, agents can choose a mix of savings and maintenance such that a higher capital stock is associated with a higher level of environmental quality.

Under their framework, it is also possible for some environmental problems to improve at low-income levels, whereas others worsen even in rich economies. For example, in the case of water quality, returns to maintenance are high and agents may be willing to sacrifice large amounts of consumption in return for improvements in quality. On the other hand, for

---

[61] The overlapping generations model is a departure from the normal utility framework used in economics. In this approach, the utility of the current generation depends on that of future generations.

other pollutants (e.g., CO_2), returns to maintenance may be low and agents may value environmental quality relatively less.

Production/consumption models

Pollution can arise from consumption and production of goods and services or the use of environmental inputs in either of these activities. Lopez (1994) presents a simple model that consists of two production sectors and assumes constant returns to scale, exogenous prices[62] and technological progress, among other things. In this model, when private producers consider only their marginal costs (MPC) (i.e., do not pay for pollution), increased output levels lead to increase in pollution levels regardless of technological progress and preferences. However, when producers pay the marginal social cost (MSC), that is, MPC plus the price of pollution, then the relationship between output and pollution levels depends on preferences and technology. If it is assumed that preferences are non-homothetic, as is more likely to be the case (Pollak and Wales, 1992),[63] the change in pollution, with increasing output, depends on the elasticity of substitution in production between pollution and other inputs, as well as the degree of relative risk aversion. In this case, the degree of relative risk aversion is defined as the rate at which consumers' marginal utility declines as they increase their consumption of goods and services. For certain plausible values of these two parameters (i.e., elasticity of substitution and risk aversion) pollution levels may rise at low-income levels and decline at high income levels, leading to the inverse U shape.

McConnell (1997) has proposed an EKC model in which pollution is generated by consumption but reduced by abatement. In this model, utility is defined as a function of consumption (C) and pollution (P). That is,

$$U = U(C, P) \tag{10.1}$$

where pollution is a function of consumption and abatement (A). That is,

---

[62] Constant returns to scale means that a 1 percent increase in inputs results in a 1 percent increase in output; exogenous prices mean that prices are given, i.e., determined from outside the model.

[63] The property of 'homotheticity' is related to constant returns to scale (see previous footnote). It is related to the fact that a set of inputs is used in the same proportions at any given level of output.

$$P = P(C, A) \tag{10.2}$$

It is assumed that output is equal to consumption plus abatement. That is,

$$Y = C + A \tag{10.3}$$

McConnell (1997) assumes that a social planner maximises Equation (10.1) subject to the constraint, Equation (10.3). He goes on to demonstrate conditions under which an inverted U curve may or may not be generated by changes in the sign of the income elasticity of demand for environmental quality. McConnell demonstrates that it is possible for pollution to decline with a zero income elasticity of demand for environmental quality, or to increase with a high-income elasticity of demand for environmental quality.

Antle and Heidebrink (1995) have used the concept of the **production possibilities frontier**[64] (PPF) to explain why economic growth and environmental quality may not necessarily be mutually exclusive. They define an economy that produces two goods, market goods (x) and environmental goods and services (e). The production functions are defined as:

$$x = x(L, K, Z) \tag{10.4}$$
$$e = e(E) \tag{10.5}$$

where L is labour and other variable inputs used to produce market goods; K is environmental capital stock used to produce x; Z is conventional capital stock (e.g., structures and machinery) and E is environmental capital stock (e.g., forests, soil, and water) used to produce environmental services. Both production functions are assumed to be concave and the usual neoclassical assumptions (i.e., perfect competition and constant returns to scale) also apply. The marginal product of the conventional capital stock is assumed to be negative at high levels of utilisation, but that of environmental capital stock is assumed to be positive. The **marginal rate of transformation** from e to x, $MRT_{e,x}$ is defined as:

---

[64] The production possibilities frontier is a function which indicates the maximum output that can be obtained with different combinations of inputs.

$$MRT_{e,x} = \frac{MP_e}{MP_x} \qquad (10.6)$$

Based on neoclassical economic theory, equilibrium is achieved where $MRT_{e,x}$ is equal to the relative prices of e and x (see Figure 10.9). Assuming e is a pure public good, which is often the case with environmental goods and services, its market value will be zero or low and therefore the price line will be steep or close to infinity. Equilibrium will thus occur at point c, with e^* units of the environmental good and x^* units of the market good being produced. The 'rational' producer will not produce below e^* units, where there are diminishing returns. That is, even though environmental goods are zero priced, there is no economic incentive for the market economy to totally exhaust the environmental capital stock. However, over time, as more environmental inputs are used up, the envrionmental capital stock declines. In response to this change, the shadow price of environmental goods rises to reverse the decline.

Figure 10.9 Production possibilities frontier for market and environmental goods

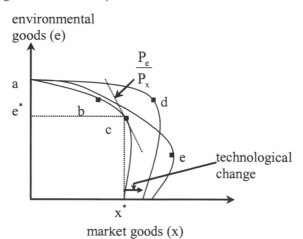

Source: adapted from Antle and Heidebrink (1995).

Technological change can shift the PPF outward. The main implication of this theory is that even if environmental quality is a pure public good, economic growth and environmental quality improvement are not necessarily mutually exclusive, given the possibility of technical change. In this framework, it is possible for public policies (e.g., environmental regulations) to move the economy to point d, where more market goods and more environmental services are produced. However, it is also possible for some government policies (e.g., large-scale dam projects) to move the economy to points such as e where the level of environmental services is reduced.

Political economy models

Many EKC studies model pollution levels as a function of per capita income without specifying the links between these two variables. According to Grossman and Krueger (1995), the strongest link is an induced policy response, which is, in turn, induced by popular demand. To quote Grossman and Krueger,

> "As nations or regions experience greater prosperity, their citizens demand that more attention be paid to the non-economic aspects of their living conditions. The richer countries tend to have relatively cleaner urban air and relatively more stringent environmental standards and stricter enforcement of their environmental laws than middle-income and poorer countries" (Grossman and Krueger, 1995:372).

Given that most environmental goods and services display public goods characteristics, the issue of market failure needs to be addressed in efforts to solve environmental problems. According to the political economy approach, as a country's per capita income increases, it becomes better able to address the issue of market failure through the political process.

To conclude this section, the point needs to be made that there is not much difficulty in constructing a model that would generate EKC-type characteristics. The challenge is to find empirical evidence that backs up the theory. In the following section, we will review a selection of empirical studies that have attempted to either support or refute the EKC hypothesis.

10.5.2 Review of Empirical EKC Studies

Grossman and Krueger (1991) conducted the path-breaking study in the EKC literature as part of a wider study to assess the environmental impacts of the North American Free Trade Agreement (NAFTA). The study utilised panel data from the Global Environmental Monitoring System (GEMS) project to estimate EKCs for SO_2, dark matter, and SPM for a number of cities worldwide.[65] Regression variables including a cubic function of per capita GDP (PPP dollars), a time trend and trade intensity were used. The analysis confirmed that ambient concentrations of SO_2 and dark matter exhibit EKCs with turning points lying between $4,000 and $5,000 in 1985 US dollars (see Table 10.3). Although they found that economic growth at middle income levels was associated with improved environmental quality, growth at higher income would be detrimental.

An EKC study conducted by Shafik and Bandyopadhyay (1992) was used as a background study for the 1992 World Development Report (World Bank, 1992). They estimated EKCs for nine different indicators of environmental quality: SPM, ambient SO_2, deforestation, lack of clean water, lack of urban sanitation, dissolved oxygen in rivers, faecal coliforms in rivers, municipal waste per capita, and CO_2 emissions per capita. They carried out panel and cross section regressions using data from 149 countries for the period 1960–1990. The two air pollutants, SO_2 and SPM were found to conform to the EKC hypothesis with turning points at $3,700 and $3,300 respectively (Table 10.3). Deforestation was not significantly related to income, while river quality worsened as income increased. Finally, both CO_2 emissions per capita and municipal waste increased significantly with increase in income.

Panayotou (1993, 1995) conducted cross-sectional EKC regressions using data for 1987/88 for 55 developing countries. All the indicators, including emissions per capita for SO_2, SPM, NO_X and deforestation, were found to conform to the inverted U shape. The turning points were $3,000, $4,500, $5,500, and $823, respectively (Table 10.3). Panayotou updated his study in 1997 by explicitly accounting for the underlying determinants of environmental quality. In order to gain a more comprehensive understanding of the income-environment relationship, he incorporated a variety of policy instruments into the analysis. This involved 'decomposing' the structural economic factors influencing the emissions of SO_2 into its pure income,

---

[65] The GEMS is a joint project of the World Health Organisation and the United Nations Environmental Program.

scale and sectoral composition effects, as well as testing independently for the impacts of the rate of growth and a policy variable. The results indicated that both the growth rate and the policy variables were highly significant, with a turning point of just below $5,000.

Table 10.3 A selection of EKC studies[a]

Study	Environmental Indicator	Turning Point (US$ per capita)	Countries/Cities/ Time Period
1. Grossman and Krueger (1991)	SO_2 Dark matter	4,000–5,000	Cities worldwide
2. Shafik and Bandyopadhyay (1992)	SO_2 SPM Deforestation	3,000–4,000	149 countries, 1960–90
3. Panayotou (1993, 1995)	SO_2 NO_x SPM Deforestation	3,000 5,000 4,500 823	55 developed and developing countries, 1987–88
4. Panayotou (1997)	SO_2	5,000	Cities in 30 developed and developing countries, 1982–84
5. Cole *et al.* (1997)	Carbon monoxide Total energy use CFCs and halons	25,100 22,500 15,400	7 regions, 1960–91, 24 OECD countries, 1970–90, 38 countries, 1986, 1990
6. Cropper and Griffiths (1994)	Deforestation	4,760 5,420	64 countries in Africa, Latin America
7. Antle and Heidebrink (1995)	Afforestation National parks	2,000 1,200	93 countries, cross-section, 1985 82 countries, cross-section, 1985
8. Asafu-Adjaye (1998)	Reforestation	5,000	83 countries, cross-section, 1985
9. Wackernagel *et al.* (1997)	Ecological footprint	21,587	52 countries, 1985
10. Asafu-Adjaye (1999)	Energy use	22,218	4 Asian countries, pooled cross section-time series, 1971–95

Notes:
[a]
SO_2 = sulphur dioxide
SPM = suspended particulate matter
NO_x = nitrous oxide
CO_2 = carbon dioxide
CFC = chlorofluorocarbons

Grossman (1993) undertook a comprehensive EKC study using data from up to 488 monitoring stations from 64 countries for the period 1977 to 1990. U-shape relationships were observed for SPM, NO_2, CO, faecalcoliform, biological oxygen demand (BOD)[66] and chemical oxygen demand (COD)[67] with turning points of $16,000, $18,500, $22,800, $8,500, $10,000 and $10,000 respectively.

In a departure from the aforementioned studies, Cropper and Griffiths (1994) estimated EKCs for deforestation for African, Latin American and Asian countries using time series data for a 30-year period. Deforestation in African and Latin American countries displayed inverted U shapes with turning points of $4,760 and $5,420 respectively. However, there were no significant relationships for Asian countries.

More recently, de Bruyn (1997) used decomposition analysis[68] to determine whether structural change or technological innovation was a major factor in the decline in SO_2 emissions in the Netherlands and West Germany during the 1980s. Although he failed to find evidence supporting the significance of structural change, he found that environmental policy fostered by international agreements provides a better explanation of why pollution tends to decline as income levels increase. It is significant to note that de Bruyn (1997) found income to be only a minor determinant of environmental policy.

Cole *et al.* (1997) used more recent data encompassing a wider set of environmental indicators including CFCs, halons, methane, nitrates, municipal waste, carbon monoxide, energy consumption and traffic volumes. They found the turning points for per capita emissions of total NO_x, SPM, and carbon monoxide to be comparable to earlier studies, implying that per capita emissions of these pollutants are beginning to decline in many advanced economies (Table 10.3). However, both CO_2 and energy use were found to increase monotonically with income. CFC's and halons, although predicted to follow a similar path as CO_2 in 1990, were found to have flattened out and decreased slightly.

---

[66] BOD is the amount of natural oxidation that occurs in a sample of water in a given time period.

[67] COD is the amount of oxygen consumed when a chemical oxidant is added to a sample of water.

[68] de Bruyn (1997) and Panayotou (1997) both advocate decomposition analysis as a preferred alternative to the reduced-form approach. This is because an expansion of the reduced-form model to include further explanatory variables increases the possibility of serious multicollinearity problems.

Other environmental indicators for which turning points were reported (Table 10.3) were afforestation/reforestation (Antle and Heidebrink, 1995; Asafu-Adjaye, 1998), and ecological footprint (Wackernagel *et al.*, 1997). Wackernagel *et al.* estimated the EKC for a sample of 52 countries using 'ecological footprints' as an indicator. The ecological footprint estimates the land and water required to sustainably provide for the average per capita consumption in each country, including the following: food, wood, energy and built area. Using a quadratic specification, the estimated turning point of $21,587 was outside the data range and the log-quadratic specification did not indicate a turning point.

Some studies have attempted to investigate the effect of political factors on pollution. However, the evidence so far has been inconclusive or contradictory. For example, Shafik and Bandyopadhyay (1992) tested for the influence of political and civil rights[69] on concentrations of various air pollutants (including SO_2) and found evidence that air quality is worse in more democratic countries. On the other hand, Torras and Boyce (1998) found evidence to support the view that less 'power-equal' countries (both with respect to democracy and income equality) have higher SO_2 emissions.

Whereas earlier studies (e.g., Shafik and Bandyopadhyay 1992; Grossman, 1993; Panayotou, 1995) showed evidence of a U-curve for SO_2, more recent research has cast doubt on an EKC for SO_2 (Stern and Common, 2001) and other air pollutants (Harbaugh *et al.*, 2000). In their study which used a much larger sample of countries over a longer period of time than previous sulfur EKC studies, Stern and Common (2001) found that the turning point estimates were sensitive to sample choice. For example, using a sample of 23 OECD countries and a random effects model, they obtained an inverted-U shape with a turning point of US$9,239 which was well within the sample range. However, using a global sample, they obtained a very high turning point of US$101,166, implying that the EKC is effectively a monotonic function of income. This finding is consistent with that of List and Gallet (1999) who also found a very high turning point for sulphur for US states when they used a long time series (1929–1994) and a wide income range (US$1,162–US$22,462).

Harbaugh *et al.* (2000) re-examined the empirical evidence for the EKC for SO_2, smoke, and total suspended particulates using data from World

---

[69] The political liberties index measures rights such as free elections, the existence of multiple parties, and decentralisation of power. The civil liberties index measures freedom to express opinions without fear of reprisal.

Bank (1992) and Grossman and Krueger (1995), with the benefit of an additional ten years of data. They also tested the sensitivity of the EKC relationship to different functional forms and econometric specifications, to the inclusion of additional covariates besides income and to the nations, cities and years sampled. They found that the location of the turning points, as well as their very existence, was sensitive to both slight variations in the data and to the econometric specification. For example, merely cleaning up or updating the original data caused the inverted-U shape to disappear. On the basis of these results, they concluded that there is little if any empirical support for the existence of an EKC for these pollutants.

10.5.3 A Critique of EKC Studies

In view of the nature and intensity of the debate about economic growth and environmental degradation, it is not surprising that the EKC hypothesis has come under heavy attack from both economists and non-economists alike. The EKC seems to suggest that countries can simply 'grow out' of any limitations brought about by the depletion of natural resources and increased environmental degradation. This view was put even more forcefully by Beckerman (1992) who said, *inter alia*, that 'the best—and probably the only way to attain a decent environment in most countries is to become rich'.

Eminent scholars such as Arrow *et al.* (1995), Rothman (1997), and Stern *et al.* (1996) have critiqued the EKC. Special issues of the following journals have been devoted entirely to the subject: *Ecological Economics* (1995), *Ecological Applications* (1996) and *Environment and Development Economics* (1996). A panel of economists, led by Kenneth Arrow, met in Sweden to consider the relationships between economic growth and environmental quality. They concluded, *inter alia*, that an inverted U-curve does not constitute evidence that environmental quality will improve in all cases or that it will improve in time to avert the adverse impacts of economic growth (Arrow *et al.*, 1995). They argued that in most cases where emissions have declined with increasing income, the reductions have been due to local institutional reforms such as environmental legislation and market-based incentives, although such reforms have tended to overlook international and intergenerational consequences.

From the current empirical evidence, it is unclear whether the EKC is the result of economic growth and therefore best tied to income increases, or whether it is merely a symptom of other underlying exogenous changes. Consequently, some of the more recent studies have attempted to find alternative approaches for analysing EKC relationships. For example, Unruh and Moomaw (1997) and Moomaw and Unruh (1997) have disputed the EKC conclusion that increase in income results in emissions reduction. Instead, they argue that reductions in emissions are triggered by specific historic events such as the 1973 oil crisis. They find that the transition to lower per capita CO_2 emissions can occur at different income levels and can occur rather abruptly. They demonstrate that, in the case of countries as different as Spain and the United States, the transitions occurred soon after the oil price shocks of the 1970s.

McConnell (1997) has suggested that although income is an important factor in explaining the EKC type relationship, other factors such as abatement costs, or the impact of pollution on production may over ride the effects of income. He therefore investigates the relationship between the demand for environmental quality and income by considering the combined effect of preferences, increasing costs of pollution abatement and the marginal utility of consumption in a growing economy in order to decompose the reduced form effect of income on pollution. He concludes that a high-income elasticity of demand for environmental quality is neither necessary nor sufficient to yield an EKC-type relationship.

Some studies (e.g., Liddle, 1996; Asafu-Adjaye, 1999) have found no evidence of trade playing a major role in determining the EKC. On the other hand others (e.g., Suri and Chapman, 1997) find the opposite result. Other studies such as Suri and Chapman (1997) have attempted to estimate the effect of economic structure or industrial organisation on the EKC relationship. Westbrook (1995) estimated the EKC for 56 developing countries including variables to represent industrial structure—defined as the shares of agriculture and services in GNP, respectively, relative to the industrial sector. He found industrial structure to be a significant factor explaining the U-curve relationship. Finally, the EKC concept is totally inadequate to explain the decline in environmental indicators such as biodiversity. This issue is considered in a little detail in the next section.

10.6   Economic Growth and Biodiversity[70]

The majority of EKC studies focus on aspects of environmental degradation such as air/water pollution and deforestation. Biodiversity belongs to a special class of environmental degradation because it involves complex ecosystems the loss of which cannot be recovered by technological advances. As such they differ from other types of environmental degradation such as pollution and deforestation for which improvements are possible to some extent. Furthermore, biodiversity levels are not related to energy use unlike pollutants commonly used in EKC studies. Thus, at the global level, there cannot be a turning point in the relationship as income increases.

Biodiversity is a complex variable that is difficult to capture with a single indicator. A naïve model of biodiversity is based on the ecological theory of island biogeography (MacArthur and Wilson, 1967) which represents the number of species (*S*) as a function of area (*A*) as follows:

$$S = CA^z \qquad (10.7)$$

where *C* and *z* are positive parameters, with *z* ranging from 0.10 to 0.35 (Wilson, 1992). Given that Equation 10.7 is non-linear in both the parameters and variables, we can express it in logarithms as follows:

$$ln\text{S} = ln\text{C(X)} + z\, ln(\text{A}) \qquad (10.8)$$

where *X* is the vector of variables whose impact on biodiversity we wish to investigate.

In a study conducted by Asafu-Adjaye (2003), the following indicators of species diversity were used as proxies for biodiversity: (1) number of known mammal species/10,000 sq km (*MAMMALS*); (2) number of known bird species/10,000 sq km (*BIRDS*); (3) number of known higher plant species/10,000 sq km (*PLANTS*); (4) percentage of bird and mammal species threatened with extinction (*PBMT*); and (4) average annual percentage change in the number of known mammal species for the period 1989/1999 (*PCMAM*).

Figure 10.10 shows graphical plots of the relationship between income (represented by GDP per capita) and the number of known mammal and bird species, respectively, for a cross-section of 100 countries. For each indicator,

---

[70] The material in this section draws heavily from Asafu-Adjaye (2003).

Figure 10.10 Plots of the relationship between income and biodiversty levels

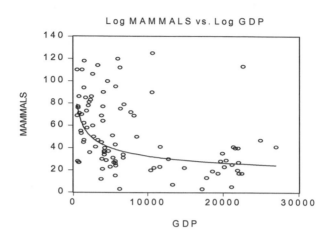

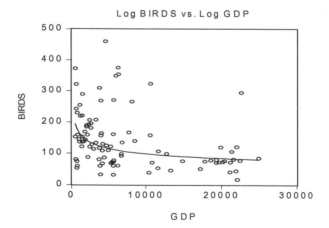

Source: Asafu-Adjaye (2003).

the plots show wide variations in species diversity especially for countries in
the less than US$10,000 category. However, ignoring the outliers in the
samples, it can be seen that there is an overall negative relationship between

income and biodiversity levels for the numbers of known mammal, bird and higher plant species, and a positive relationship between income and the percentage of birds and mammals threatened with extinction. Thus, an EKC type relationship can be ruled out for biodiversity.

To quantify these relationships more precisely and examine the impact of economic growth on biodiversity, the following regression models were estimated:

$$lnE_{ij} = \alpha_0 + \alpha_1 lnGDP_i + \alpha_2 lnAGRICPC + \alpha_3 lnFREE_i + \alpha_4 lnFOREX_i +$$
$$\alpha_5 lnPOPDENS_i + \alpha_6 lnPDLAND_i + \alpha_7 lnPPLAND$$
$$+ \alpha_8 lnCLIMATE_i + \varepsilon_i \qquad (10.9)$$

For indicators 4 and 5 the following regression model was estimated:

$$E_{ij} = \alpha_0 + \alpha_1 lnGDP_i + \alpha_2 lnAGRICPC + \alpha_3 lnFREE_i + \alpha_4 lnFOREX_i$$
$$+ \alpha_5 lnUPOPGRO + \alpha_6 lnPDLAND_i + \alpha_7 lnPPLAND$$
$$+ \alpha_8 lnCLIMATE_i + \varepsilon_i \qquad (10.10)$$

where:

E_{ij} = indicator of biodiversity level for country i, j=1 (mammals), 2 = (birds), 3 = (plants), 4 = (pbmt) and 5 (pcmam);

GDP_i = GDP per capita (in PPP terms, 1995 International $)

$AGRICPC_i$ = agricultural value added as a percentage of GDP;

$FREE_i$ = index of economic freedom, 0 (most free) to 20 (least free);

$FOREX_i$ = black market premium on exchange rates, 0 = premium of 210 percent or more, 10 = premium of 0 percent;

$POPDENS_i$ = population density;

$UPOPGRO_i$ = urban population growth;

$PDLAND_i$ = percentage of land developed for agriculture and other uses;

$PPLAND_i$ = percentage of protected land area; and

$CLIMATE_i$ = dummy variable for climate where: (1) cold and cold temperate countries; (2) sub-tropical and dry countries; and (3) wet tropical countries; and

ε_i = a random error term.

Gross domestic product was included in the model because the level of biodiversity is directly related to the level of economic output. *AGRICPC* was included to test the hypothesis that biodiversity decline is not only affected by the level of economic activity but also by the composition of economic output. It was hypothesized that a well developed economic, social and political institutions that help to internalise the value of biodiversity into decision-making processes of the state and individuals.

The variable *FREE*, an index of economic freedom, was used as a proxy for institutional development in a country. This index is published by the Fraser Institute (Gwartney and Lawson, 1997) and ranges from 0 (most free) to 20 (least free). It is expected that a negative relationship will exist between the degree of economic freedom and biodiversity level.

The variable *FOREX*, an index of the black market premium on foreign exchange, was used to represent the macroeconomic environment in a country. It is expected that macroeconomic policies will affect the level of environmental degradation, which in turn will affect biodiversity resources. For example, a high black market exchange rate indicates a restrictive trade policy stance and overvaluation of the domestic currency that could negatively affect private rents from timber exports. However, on the other hand, it could also discourage development of non-timber forest product industries. *FOREX* ranges from 0 to10, with 0 representing a black market exchange rate premium of 210 percent or more and 10 representing a premium of 0 percent. The variables *PDLAND* and *PPLAND* were used as proxies for habitat size, while *POPDENS* and *UPOPGRO* were used to represent population pressure. Finally, a dummy variable was included to account for climatic effects on biodiversity. It is a well-known fact that species diversity increases as one moves from the polar areas towards the equator. Studies (e.g., Holm-Hansen *et al.*, 1993; Termura *et al.*, 1990) suggest that phenomena such as ozone depletion and CO_2 emissions that contribute to global warming may indirectly increase biodiversity loss.

Data on GDP per capita, the Index of Economic Freedom, and the black market exchange rate premium were obtained from Gwartney and Lawson (1997), while data on population density and the percentage of agricultural value added in GDP were taken from World Bank (1999). Data on the remaining variables were obtained from *World Resources* (WRI, 1990, 1999). The list of countries included in the sample is given in the Appendix 10.1.

The results (see Tables 10.4 and 10.5) indicate the following:

1. The level of economic activity income has a significant negative effect on species density for mammals and birds but not for higher

plants. It also appears to have an adverse effect on the average annual percentage change in the number of known mammal species.

2. The proxy for the composition of economic output (*AGRICPC*) is highly significant for the average annual percentage change in the number of known mammal species. This result provides some justification for the view that it is not only the level of economic output *per se* which is injurious to biodiversity, but also the composition of that output. This confirms the fact that conversion of habitat for agricultural and other purposes is one of the major threats to biodiversity conservation.

Table 10.4 Regression estimates for determinants of biodiversity
(numbers of mammals and birds)

Variable	No. of mammals /10,000 sq km[a]		No. of birds /10,000 sq km[a]	
	Coefficient	t-Statistic	Coefficient	t-Statistic
Intercept	6.00***	2.98	4.02***	3.22
*ln*GDP	−0.28**	−2.26	−0.04*	−1.37
*ln*AGRICPC	−0.08	−0.60	0.09	0.94
*ln*FREE	−0.10	−0.41	0.08	0.43
*ln*FOREX	0.01*	1.87	0.02	0.84
lnPOPDENS	−0.10**	−1.99	−0.11***	−2.85
*ln*UPOPGRO	-	-	-	-
*ln*PPLAND	0.11***	2.31	0.12***	2.91
*ln*PDLAND	0.02	0.30	0.03	0.49
*ln*CLIMATE	0.54***	2.64	0.48***	3.25
R^2	0.34		0.35	
Adjusted R^2	0.29		0.29	
Std error	0.64		0.53	
F-statistic	5.95***		5.89***	
N	99		98	

Notes:
[a] The dependent variable is the logarithm of number/10,000 sq km.
***, **, and, * indicate statistical significance at the 1%, 5% and 10% level, respectively, for a one-tailed test.

Source: Asafu-Adjaye (2003).

Table 10.5 Regression estimates for determinants of biodiversity (number of higher plants and change in number of mammals)

Variable	No. of higher plants /10,000 sq km[a]		Average % change in no. of mammals 1989–99[b]	
	Coefficient	t-Statistic	Coefficient	t-Statistic
Intercept	3.51*	1.45	22.88**	2.14
lnGDP	0.11	0.56	−1.40*	−1.51
lnAGRICPC	−0.11	−0.63	−2.70***	−2.65
lnFREE	0.27	0.70	−1.91*	−1.30
lnFOREX	0.30***	6.01	−0.06	−0.20
lnPOPDENS	−0.20***	−2.61	-	-
lnUPOPGRO	-	-	−0.02	−0.05
lnPPLAND	0.10*	1.35	−0.12	−0.39
lnPDLAND	0.19	1.22	−1.04**	−2.04
lnCLIMATE	1.39***	4.88	5.85***	3.64
R^2	0.44		0.28	
Adjusted R^2	0.39		0.18	
Std error	1.02		3.63	
F-statistic	8.63***		2.91***	
N	98		71	

Notes:

[a] The dependent variable is the logarithm of number/10,000 sq km.

[b] The dependent variable is the logarithm of the average annual percentage change in the number of known mammal species from 1989 to 1999.

***, **, and, * indicate statistical significance at the 1%, 5% and 10% level, respectively, for a one-tailed test.

Source: Asafu-Adjaye (2003).

3. The black market exchange rate premium (*FOREX*) is positive and statistically significant for mammals and higher plants, with the effect being relatively stronger for the latter. This particular variable is an indicator of distortions in the economy and the implication here is that removal of such distortions could lead to an improvement in not only the economy but also aspects of the environment.

4. As expected, population density has a highly significant negative effect on biodiversity loss in general.

5. *PPLAND* is significant for all the three indicators of species density, while *PDLAND* is significant for the average annual percentage change in the number of known mammal species. These results lend empirical support for the view that space is a limiting factor to biodiversity protection, and provide a justification for the policy of setting aside nature conservation areas.

6. The coefficients on climate are significant for all the indicators of biodiversity, confirming that climate contributes to biodiversity decline.

10.7 Conclusions and Policy Implications

In the last few years, progress has been made in enhancing our understanding of the relationship between the environment and development. On the basis of current research, a number of conclusions can be reached.

1. The so-called U curve relationship between environmental quality and income does not apply to all environmental indicators. The review conducted in this chapter demonstrates that the indicator which consistently displays such a relationship, based on the number of studies confirming it, is SO_2. However, even this has been disputed on a number of grounds. A few studies have found the EKC hypothesis to hold for SPM, deforestation, afforestation and energy use. However, these results cannot be generalised.

2. The EKC is an inappropriate model to describe the relationship between economic growth and the environment in the case of environmental indicators such as biodiversity. Biodiversity involves the irreversible loss of valuable ecosystems and the current evidence suggests that economic growth contributes significantly to its decline.

3. Even where a U curve relationship is found, the turning points tend to be much higher than the per capita incomes of the countries

involved. For example, in Asafu-Adjaye (1999) the turning point for energy was US$22,218 for a sample of four countries comprising, India, Indonesia, South Korea and Japan. However, real per capita income for India, the least well off in the sample based on GDP, is US$439. These results imply that, for many developing countries, environmental problems may worsen in the foreseeable future.

4. It has been suggested that trade openness could help in reducing environmental pollution. However, there is no significant evidence in the literature to support this proposition.

5. Many of the studies that have found a U-shaped relationship between economic growth and the environment do not convincingly explain how growth affects the environment. The empirical evidence supporting the existence of an EKC relationship is not clear-cut. This implies that the relationship between the environment and development is too complex to be adequately represented by simple economic models.

A general conclusion that can be reached is that developing countries will not automatically grow out of their environmental problems. However, economic growth is necessary for developing countries to make a dent in their environmental problems. The fact that many countries' per capita incomes are below the potential turning points supports the view that restricting economic growth to save the environment may not be a socially optimal decision. Progress towards the turning point could be boosted by a combination of prudent economic policies and environmental regulations.

Some of the studies reviewed above indicate that institutional factors do exert a significant influence on the relationship between income and environmental degradation. Barbier (1997) argues that policies that aim to improve the operation of markets are more likely to reduce the 'environmental price' of economic growth, thereby flattening out the income-environment relationship and possibly achieving a lower turning point. Such policies include the removal of distortionary subsidies, introduction of more secure property rights over resources and implementation of economic instruments to internalise externalities. Barbier's argument is reinforced to some extent by Asafu-Adjaye (2003) finding that improvement in economic freedoms (a proxy for the level of institutional development) is associated with improvement in the numbers of mammal and bird species threatened with extinction.

In addition to not providing avenues for economic agents to internalise environmental costs, a poorly developed institutional framework also affects the government's ability to monitor environmental degradation and/or pollution which, in turn, retards the development of effective environmental policies. In his EKC study of SO_2 emissions, Panayotou (1997) found that improved policies and institutions in the form of more secure property rights, better enforcement of contracts and effective environmental regulation could assist in making the EKC flatter. For example, reductions in SO_2 emissions in the Netherlands and West Germany have been brought about by tougher environmental regulations requiring better end-of-pipe technology. Komen and others' research indicates that increased public spending on environmental research and development, while possibly leading to environmental improvements, may also act as a catalyst for private investment in cleaner technologies (Komen *et al.*, 1997).

In the following sections, we look at policy responses to the environmental degradation problem at the local/national and global levels.

10.7.1 National and Local Level Policy Response

At the local and national levels, there are a variety of options for dealing with the problem of environmental degradation. In Chapter 4, we discussed a number of market-based instruments including standards, taxes, subsidies, and marketable permits. In this section, we will take a brief look at another set of options for tackling the pollution problem—**voluntary incentives**.

Most pollution policy instruments rely on coercion (e.g., fees or penalties) or some form of financial incentives. As the name suggests voluntary incentives rely on voluntarism and self-regulation. There are two major categories of voluntary incentive mechanisms—**voluntary agreements** and **voluntary, incentive and community-assistance** programs.

Voluntary agreements

A voluntary agreement (VA) is basically a 'contract' between a government agency and industry in which environmental goals and deadlines have been negotiated and subsequently agreed upon by both parties (Barde, 1995; Carraro and Sinisalco, 1996). In a VA, the industry is self-committed to taking appropriate measures to meet these goals. In the event of non-compliance there are no real sanctions, except that regulations and coercive measures may be imposed at the end of the contract period. One important feature of a VA is that although pollution levels may be fixed within a geographical area, the industry is free to pursue the most cost-effective measures to achieve the agreed objective. One advantage of VAs is that they can be combined with regulatory requirements in two different ways: (i) they can be implemented before any subsequent regulations; and (ii) they can reinforce existing regulations that may be poorly enforced.

Like most other policy instruments there are arguments for and against VAs. For example, it has been argued that voluntary agreements reduce government control over industry, or that they could encourage 'collusion' between government and industry. In practical terms, voluntary agreements may become difficult to manage when there are many sources requiring regulations, and when pollution spillovers affect communities who are not party to the voluntary agreement. However, on the positive side, voluntary agreements are flexible, transparent and could provide incentives for technological innovation.

Voluntary, incentive and community-assistance programs

Voluntary, incentive and community-assistance programs depend on the commitment, enthusiasm and goodwill of local community groups to undertake conservation projects. As the name suggests, this approach is purely voluntary, although, in some cases, grants may be advanced to facilitate the initiatives. This approach has been recommended as a mechanism for conserving biodiversity on private property (OECD, 1996). The main attraction of such programs is that they are non-interventionist in nature and require minimal administrative costs. A disadvantage is that they can be difficult to target and monitor without incurring high administrative costs. These sorts of schemes work best in cases where the participants have a genuine interest in the goal of the project.

10.7.2 Global Policy Response

If the EKC relationship is true, then countries in early stages of development will, as a result of lack of access to technology and capital, adopt industrial processes that are polluting, or will mine their natural resources. In this respect, it is self-serving for the rich nations to demand that the poorer nations cut back on pollution or natural resource use. Therefore, it has been proposed that the rich nations could compensate the poorer ones for forgoing income from natural resource exploitation. In the case of global public goods such as biodiversity resources, individuals and countries have no economic incentive to invest in them. Thus, there is the need for more international initiatives that assist the developing countries to improve the management of their biodiversity resources. Various forms of transfer mechanisms have been proposed including **international financing** of conservation projects and **debt-for-nature swaps**.

International financing of conservation projects

A number of initiatives by which advanced countries can assist developing countries to reduce environmental degradation have been proposed. These include using savings from military expenditure made since the Cold war to retire debt in developing countries and to fund conservation projects, and flexible interest repayment terms on debt for sustainable development projects. Another initiative already underway is the Global Environmental Facility (GEF). The GEF was established by the World Bank and the United Nations Environment Programme. The GEF aims to provide concessional loans to developing countries for projects associated with protection of the ozone layer, reduction of greenhouse emissions, protection of international water resources, and protection of biodiversity

Specific projects that are eligible for assistance under the scheme include development of alternative energy sources, afforestation, conservation of tropical forests and investments to prevent oil spills and ocean pollution. Although the scheme is a worthwhile one, some countries have complained about inadequate funds. It would appear that the issue of financing the recurrent expenditure for the fund has not been properly worked out.

Debt-for-nature swaps

The concept of debt-for-nature swaps has been proposed as a way of assisting developing countries to protect their environment and at the same time reduce their foreign debt. The idea originated from the concept of **debt conversion**. After Mexico's 1982 financial crisis, it threatened to default on its foreign debt. This threat forced international banks to accept the fact that some debt-ridden countries were incapable of fully repaying their debts. Consequently, banks began to minimise their risk by selling high-risk debts to other banks at discounted values. Such debts became known as secondary debts or loans, and soon a market for these loans began to develop. A typical debt conversion consists of swapping secondary debt for equity in the debtor country and offers a low-cost way of investing in a developing country. For example, in 1986, Chrysler Motor Company bought off Mexico's foreign debt with a face value of about US$108 million for US$65 million. In return, the Mexican government provided about US$100 million in pesos for the construction of a car manufacturing plant.

Debt-for-nature swaps[71] operate along the same lines as debt-for-equity swaps, except that, as the name suggests, the investments are targeted at preserving the environment. For example, the US environmental lobby group, Nature Conservancy bought about US$2.2 million of Brazil's foreign debt for US$850,000. Most of the money has been paid into a fund to be used to conserve a tropical rain forest in the country. The World Wildlife Fund for Nature has been involved in debt-for-nature swaps in a number of developing countries. By 1991, about 19 debt-for-nature swaps worth over US$100 million had taken place. However, this amount is miniscule in comparison to total developing country debt of over US$1.3 trillion in the same period.

Although well intentioned, debt-for-nature swaps have been criticised on a number of grounds. Some people are of the view that a good proportion of foreign debt has been incurred by totalitarian or corrupt governments and that using debt-for-nature swaps to retire such debt is tantamount to legitimising illegal or immoral transactions. For example, it has been alleged that the former president of the Philippines, the late Ferdinand Marcos, incurred millions of dollars of national debt which were improperly used and as such the people of the Philippines should not be held accountable for such debts. Another argument against debt-for-nature swaps is that they interfere with national sovereignty in the sense that they allow foreigners to dictate how governments in developing countries should allocate their expenditures. It has been alleged that some of the debt-for-nature swap projects are designed more for research and exploitation of natural resources than for conservation (Mahony, 1992).

The number and size of debt-for-nature swaps, to date, have been such that none of these criticisms have been proven beyond doubt. Realistically, it is unlikely that this approach will make any significant dent in Third World debt. It would appear that the major benefit of these schemes is in raising

---

[71] See Hansen (1989) for an overview.

awareness of sustainable development issues and making some contribution towards the achievement of that goal.

10.8 Summary

This chapter has tackled the complex subject of the relationship between trade and the environment. After describing some general trends in economic and environmental indicators in the last few years, the discussion focussed on various dimensions of the trade-environment relationship, resulting in a set of policy recommendations on trade and the environment. The discussion of the trends suggested that increase in population will increase the demand for food, clothing and shelter. The consumption of resources such as energy will continue to escalate because energy is a major input in industrial production. Energy use was found to be highly correlated with economic growth and social indicators such as infant mortality and life expectancy, implying that energy is vital for development. However, to achieve sustainable development, there is a need to develop alternatives forms of energy (e.g., renewable energy).

A considerable part of the discussion centered on the relationship between trade and the environment at different stages of development or income. At issue was the environmental Kuznets U-curve hypothesis which stipulates that, as the per capita income of a country increases, environmental degradation will initially increase, but then will eventually decline once a maximum level has been reached. The maximum level of income at which degradation begins to decline is referred to as the turning point. A review of empirical studies revealed that there is mixed support for the EKC hypothesis. The EKC relationship appears to hold for some indicators such as SO_2, deforestation and SPM. In the case of CO_2, all the studies indicate that it increases with income, and has no turning point. Also, the EKC has been shown not to hold for biodiversity. Another important observation about the EKC is that in some cases where turning points have been observed, the turning point is much higher than per capita incomes of the countries involved.

In conclusion, much still remains to be accomplished in explaining why EKCs arise. In the mean time, it is clear that countries will not automatically 'grow' out of their environmental problems. The role of institutions in countries that have obtained low levels of environmental pollution was found to be significant. Policy responses to environmental degradation were discussed, with emphasis on voluntary incentive mechanisms. In general, these approaches could offer a cost-effective and flexible means of reducing local and global pollution.

Key Terms and Concepts

biodiversity
biodiversity indicators
black market exchange rate premium
comparative advantage
debt conversion
debt-for-nature swaps
environmental Kuznets curve
environmental quality
environmental transition hypothesis
turning point

index of economic freedom
island biogeography theory
marginal rate of transformation
overlapping-generations model
political economy model
production possibilities frontier
production/consumption model of pollution
voluntary agreements
voluntary incentives

Review Questions

1. Describe what happens as natural resources are depleted.

2. Explain why energy use is positively related to a social indicator such as life expectancy.

3. Explain the theory of comparative advantage.

4. Explain the term 'turning point' which is commonly associated with the EKC hypothesis.

5. Briefly outline two theories that attempt to explain the EKC hypothesis.

6. State the advantages and disadvantages of a voluntary incentive scheme such as a voluntary agreement.

7. What are some of the factors that affect the EKC relationship?

Exercises

1. Are the conclusions of the theory of comparative advantage valid when there are environmental externalities associated with increased trade? Give practical examples to support your answer.

2. Select two EKC studies from the reference list. Read the full-length articles and provide critical summaries.

3. Critically assess the merits and demerits of debt-for-nature swaps.

4. Give possible reasons why a U-curve relationship may not necessarily mean that environmental pollution will decrease.

5. Discuss the possible effects of increased trade on biodiversity resources.

References

Andersen, K. (1992). Effects on the Environment and Welfare of Liberalising World Trade: The Cases of Coal and Food. In K. Andersen and R. Blackhurst, (eds.), *The Greening if World Trade Issues*. Harvester Wheatsheaf, London.

Antle, J.M. and Heidebrink, G. (1995). Environment and Development: Theory and International Evidence. *Economic Development and Cultural Change*, 43(3): 603-625.

Arrow et al. (1995). Economic Growth, Carrying Capacity, and the Environment. *Science*, (268):520-521.

Asafu-Adjaye, J. (2003). Biodiversity Loss and Economic Growth: A Cross Country Analysis. *Contemporary Economic Policy,* 21(2):173-185.

Asafu-Adjaye, J. (1998). An Empirical Test of the Environmental Transition Hypothesis. *Indian Journal of Quantitative Economics,* XIII: 67-91.

Asafu-Adjaye, J. (1999). The Environment and Development: Theory and Empirical Evidence. *International Journal of Development Planning Literature*, 14(1): 117-134.

Bandara, J.S. and Coxhead, I. (1999). Can Trade Liberalisation Have Environmental Benefits in Developing Country Agriculture? A Sri Lankan Case Study. *Journal of Policy Modeling*, 21(3): 349-374.

Barbier, E.B. (1997). Environmental Kuznets Curve Special Issue. *Environment and Development Economics*, 2: 369-381.

Barde, J-P. (1995) Environmental Policy and Policy Instruments. In H. Folmer, J.L. Gabel, and H. Opschoor (eds.), *Principles of Environmental and Resource Economics: A Guide for Students and Decision Makers*. Edward Elgar, Aldershot, UK.

Beghin, J., Dessus, S., Holst, R., and van der Mensbrugghe, D. (1997). The Trade and Environment Nexus in Mexican Agriculture: A General Equilibrium Analysis. *Agricultural Economics*, 17: 115-131.

Carraro, C. and Sinisalco, D. (1996) Voluntary Agreements in Environmental Policy: A Theoretical Appraisal. In A. Xepapadeas (ed.), *Economic Policy for the Environment and Natural Resources: Techniques for the Management and Control of Pollution*. Edward Elgar, Cheltenham, UK.

Chichilnisky, G. (1994). North South Trade and the Global Environment. *American Economic Review*, 84: 851-875.

Cole, M.A., Rayner, A.J. and Bates, J.M (1997). The Environmental Kuznets Curve: An Empirical Analysis. *Environment and Development Economics*, 2: 401-416.

Daly, H. and J.B. Cobb (1989). *For the Common Good: Redirecting the Economy Toward Community, the Environment, and a Sustainable Future*. Beacon Press, Boston.

Daly, H.E. (1993). The Perils of Free Trade. *Scientific American*, November.

de Bruyn, S.M. (1997). Explaining the Environmental Kuznets Curve: Structural Change and International Agreements in Reducing Sulphur Emissions. *Environment and Development Economics*, 2: 485-503.

Diamond, P. (1965). National Debt in a Neoclassical Growth Model. *American Economic Review*, 55: 1126-50.

Economic Commission for Asia and the Pacific (ESCAP) and Asian Development Bank (ADB) (1995). *State of the Environment in Asia and the Pacific*, United Nations, New York.

Grossman, G. (1993). Pollution and Growth: What Do We Know? CEPR DP-848, Centre for Economic Policy Research, London.

Grossman, G.M. and Krueger, A.B. (1995). Economic Growth and the Environment. *Quarterly Journal of Economics*, 2: 353-375.

Grossman, G.M. and Krueger, A.B. (1991). *Environmental Impacts of a North American Free Trade Agreement*. National Bureau of Economic Research Working Paper 3914, NBER, Cambridge MA.

Gwartney, J. D. and Lawson, R. A. (1997). *Economic Freedom of the World: 1997 Annual Report*. The Fraser Institute, Vancouver, Canada.

Hansen, S. (1989). Debt-for-Nature Swaps: Overview and Discussion of Key Issues. *Ecological Economics*, 1: 77-93.

Hair, J.D. (1993). GATT and the Environment. *Journal of Commerce*, December.

Harbaugh, W., Levinson, A., and Wilson, D. (2000). Re-examining the Evidence for an Environmental Kuznets Curve. NBER Working Paper No. 7711.

Holm-Hansen, O., Helbling, E.W., and Lubin, D. (1993). Ultraviolet Radiation in Antartica: Inhibition of Primary Production. *Photochemistry and Photobiology*, 58: 567-570.

John, A. and Pecchenino, R. (1994). An Overlapping Generations Model of Growth and the Environment. *Economic Journal*, 104: 1393-1410.

Komen, M.H.C., Gerking, S. and Folmer, H. (1997). Income and Environmental R&D: Empirical Evidence from OECD Countries. *Environment and Development Economics*, 2: 505-515.

Kuznets, S. (1955). Economic Growth and Income Inequality. *American Economic Review*, 49: 1-28.

Liddle, B.T. (1996). Environmental Kuznets Curves and Regional Pollution. Paper presented to the 4th Biennial Conference of the International Society for Ecological Economics, Boston University, Boston, MA.

List, A.J. and. Gallet, C.A. (1999). The Environmental Kuznets Curve: Does One Size Fit All? *Ecological Economics,* 31: 409-423.

Lopez, R. (1994). The Environment as a Factor of Production: The Effects of Economic Growth and Trade Liberalization. *Journal of Environmental Economics and Management,* 27: 163-184.

Lutz, E. (1992). Agricultural Trade Liberalisation, Price Changes, and Environmental Effects. *Environmental and Resource Economics,* 2:79-89.

MacArthur, R.H., and Wilson, E.O. (1967). *The Theory of Island Biogeography.* Princeton University Press, Princeton, NJ.

Mahony, R. (1992). Debt-for-Nature Swaps: Who Really Benefits? *Ecologist,* 22: 97-103.

McConnell, K.E. (1997). Income and the Demand for Environmental Quality'. *Environment and Development Economics,* 2: 383-399.

Moomaw, W.R. and Unruh, G.C. (1997). Are Environmental Kuznets Curves Misleading Us? The Case of CO_2 emissions. *Environment and Development Economics,* 29: 451-463.

Organisation for Economic Co-operation and Development, OECD (1996). *Making Markets Work for Biological Diversity: The Role of Economic Incentive Measures.* OECD, Paris.

Panayotou, T. (1995). Environmental Degradation at Different Stages of Economic Development. In I. Ahmed and J.A. Doeleman (eds.), *Beyond Rio: The Environmental Crisis and Sustainable Livelihoods in the Third World.* MacMillan, London.

Panayotou, T. (1997). Demystifying the Environmental Kuznets Curve: Turning a Black Box into a Policy Tool. *Environment and Development Economics,* 2: 465-484.

Panayotou, T. (1993). Empirical Tests and Policy Analysis of Environmental Degradation at Different Stages of Economic Development. Working Paper, WP238, Technology and Employment Programme, ILO, Geneva.

Pollak, R.A. and Wales, T.J. (1992). *Demand System Specification and Estimation.* Oxford University Press, New York.

Ruttan, V.W. (1971). Technology and Environment. *American Journal of Agricultural Economics*, 53: 707-17.

Samuelson, P. (1958). An Exact Consumption-Loan Model of Interest With or Without the Social Contrivance of Money. *Journal of Political Economy*, 66: 467-82.

Selden, T.M. and Song, D. (1994). Environmental Quality and Development: Is There a Kuznets Curve for Air Pollution Emissions? *Journal of Environmental Economics and Management*, 27: 147-162.

Shafik, N. and. Bandyopadhyay, S. (1992). Economic Growth and Environmental Quality: Time Series and Cross Country Evidence. Background Paper for *World Development Report*, WPS, The World Bank, Washington D.C.

Solomon, A. (1990). *Towards Ecological Sustainability in Europe: Climate, Water Resources, Soils and Biota*. IISA RR-90-6, Laxenburg, Austria.

Stern, D.I. and M.S. Common (2001). Is there an Environmental Kuznets Curve for Sulphur? *Journal of Environmental Economics and Management*, 41: 162-178.

Suri, V. and Chapman, D. (1998). Economic Growth, Trade and Energy: Implications for the Environmental Kuznets Curve. *Ecological Economics*, May, 195-208.

Temura, A.H., Sullivan, J.H. and Ziska, L.H. (1990) Interaction of Elevated UV-B Radiation and Carbon Dioxide on Productivity and Photosynthetic Characteristics in Wheat, Rice, and Soybean. *Plant Physiology*, 94: 470-475.

Unruh,G.C. and Moomaw, W.R. (1997). An Alternative Analysis of Apparent EKC-Type Transitions. *Ecological Economics*, special issue on EKC.

Wackernagel et al. (1997). *Ecological Footprints of Nations: How Much Nature do They Use? How Much Nature Do They Have?* The Earth Council, San Jose, Costa Rica, 10 March.

Wilson, E.O. (1992). *The Diversity of Life*. Harvard University Press, Cambridge, MA.

World Bank (2003). *World Development Indicators 2003*. CD-ROM Version. World Bank, Washington, D.C.

World Bank (1999). *World Development Indicators 1999*. CD-ROM version, Washington, D.C.: World Bank, 1999.

World Bank (1992). *World Development Report 1992: Development and the Environment*. Oxford University Press, New York.

World Commission on Environment and Development, WCED. (1987). *Our Common Future*. Oxford University Press, New York and Oxford.

World Resources Institute, WRI. (1999). *World Resources 1999-2000*. Oxford University Press, New York.

World Resources Institute, WRI. (1990). *World Resources 1990-1991*. Oxford University Press, New York.

Appendix 10.1 List of Countries for Study in Section 10.6

USA	Mauritius	Iran	Australia
Switzerland	Portugal	Brazil	Guatemala
Singapore	Greece	Poland	Sri Lanka
Norway	Korea	Tunisia	China
Kuwait	Chile	South Africa	Philippines
Denmark	Guinea-Bissau	Jamaica	El Salvador
Japan	Malaysia	Bulgaria	Papua New
Guinea	Dominican Republic	Mexico	Ukraine
Panama	Iceland	Nicaragua	Indonesia
Canada	Oman	Ecuador	Cameroon
Belgium	Argentina	Romania	Pakistan
Austria	Venezuela	Estonia	Ghana
France	Thailand	Namibia	Honduras
Italy	Uruguay	Lithuania	Senegal
Germany	Hungary	Paraguay	Benin
Netherlands	Colombia	Gabon	Uganda
Sweden	Fiji	Egypt	Kenya
United Kingdom	Costa Rica	Morocco	India
Finland	Syria	Peru	Congo
Ireland	Belize	Latvia	Mali
New Zealand	Algeria	Cote d'Ivoire	Madagascar
Spain	Turkey	Trinidad and Tobago	Nigeria
Bangladesh	Burundi	Sierra Leone	Rwanda
Nepal	Togo	Chad	Zambia
Malawi	Niger	Central African Republic	

11. Sustainable Development

Objectives

After studying this chapter you should be in a position to:

❑ explain the various definitions of sustainable development and 'sustainability';

❑ describe ways of measuring sustainable development;

❑ describe ways of incorporating elements of sustainability into economic decision-making; and

❑ enumerate difficulties associated with implementing sustainable development policies.

11.1   Introduction

The publication that many would agree has set the environmental agenda for the past decade and the future is the report, *Our Common Future* (WCED, 1987), which served as the background for the historic Earth Summit. The Earth Summit raised hopes and expectations about achieving sustainable development. Through Agenda 21 (see Box 11.1) and other important documents such as the Rio Declaration and the Statement of Principles for the Sustainable Management of Forests, a comprehensive environment and development agenda was drawn up for the 21st century. Unfortunately, within the past decade, progress in these areas has not matched the expectations. In many countries around the world, poverty has increased and environmental degradation has continued at an alarming pace. The World

Box 11.1 Agenda 21

Agenda 21 is an 800-page agreement consisting of 120 action programs covering nearly every aspect of all human activities that impact on the natural environment. Agenda 21 urges nations to:

> "...harmonise the various sectoral economic, social and environmental policies and plans that are operating in the country....to ensure socially responsible economic development while protecting the resource base and environment for the benefit of future generations....[National strategies] should be developed through the widest possible participation" (United Nations 1992:67).

The cost of the package is estimated at US$600 billion per annum for developing countries. The advanced countries are supposed to provide an extra US$125 billion in funding by the year 2000.

Agenda 21 comprises four basic dimensions, namely:
- Social and economic dimensions
- Conservation and management of natural resources for development
- Strengthening the role of major groups, and
- Means of implementation

It can be seen from the above that the diversity of the dimensions covered reflects the complexity of sustainable development.

Agenda 21 is a broad statement of principles and is not legally binding. Although it is a morally compelling document, it is not mandatory and the signatories can opt out if they choose to. Perhaps the greatest constraint facing the implementation of Agenda 21 is lack of financial resources. There is the danger that some countries will be selective about which aspects they implement depending on their budgetary positions. It is conceivable that the countries that are in urgent need of sustainable development may be the least able to afford it.

Source: United Nations (1992).

Summit on Sustainable Development (WSSD), which was held in Johannesburg, South Africa in August/September 2002, sought to overcome the obstacles to achieving SD. The WSSD, among other things, affirmed commitment to Agenda 21 and resolved to strengthen the concept of SD and the important linkages between poverty, the environment and the use of natural resources. Several priority areas were identified under the acronym WEHAB, which stands for water and sanitation, energy, health, agriculture, and biodiversity protection and ecosystem management.

It is often said that 'beauty lies in the eyes of the beholder', and the same can be said of sustainable development. Sustainable development (or sustainability) has been defined in several different ways. In order to make progress towards SD, it is important to clarify what the term means. This chapter canvasses a number of definitions of SD from various perspectives and then goes on to discuss the conditions for sustainable development by introducing the concepts of 'weak' and 'strong' sustainability and their implications for SD. This is followed by an outline of various indicators for assessing progress towards SD. The chapter concludes with a brief discussion of how to operationalise SD.

11.2　What is Sustainable Development?

According to Mustafa Tolba, sustainable development has become an article of faith, a 'shibboleth', often used but little explained (Tolba, 1987). Does it amount to a strategy? Does it apply only to renewable resources? What does the term actually mean? 'Sustainable development', 'sustainable economic development', or 'ecologically sustainable development' can be given various definitions depending on one's environmental ideology. In this discussion, we make a distinction between economic 'growth' and 'development'. The former can be characterised by increase in the level economic indicators such as GNP per capita, whereas the latter is a broader concept involving quality of life indicators (e.g., educational and health status). In this discussion, we consider briefly the meaning of SD from various perspectives including 'neoclassical' economics, ecology, sociology, intergenerational equity and materials balance.[72]

---

[72] In a review of sustainable development, the environmental economist John Pezzey counted over sixty different definitions (Pezzey, 1989a).

11.2.1 The Neoclassical Economics Definition

Traditional 'neoclassical' economics supposes that economic growth is maximised when all opportunities to increase the efficiency of resource use have been exhausted. In this approach, SD can be defined as the maintenance of a constant per capita consumption across all generations (Solow, 1956) or the maintenance of non-declining per capita income over the indefinite future (Pezzey 1989b). In this approach, no distinction is made between natural capital and man-made capital. It is assumed that natural, physical and human capital can be substituted for each other to a high degree. Furthermore, this definition ignores changes in natural capital stocks and environmental quality. This definition, in effect, equates SD with sustainable economic growth. However, increase in economic growth does not necessarily imply that well-being is increasing over time.

11.2.2 The Ecologists' Perspective

From the ecological perspective, the quality of life depends on environmental quality. Therefore, retaining ecological integrity and the assimilative capacity of the natural environment is crucial for the functioning of the economic system.

There have been a number of ecological definitions of SD. These include the following:

Ecologically sustainable development is a condition in which society's use of renewable resources takes place without destruction of the resources or the environmental context which they require (Solomon, 1990).

Ecologically sustainable development means using, conserving and enhancing the community's resources so that ecological processes, on which life depends, are maintained, and the total quality of life, now and in the future, can be increased (Government of Australia, 1992).

A system that is healthy and free from 'distress syndrome' if it is stable and sustainable, that is, if it is active and maintains its structure (organization) function (vigor) and autonomy over time and is resilient to stress (Costanza, 1994).

A common thread of the ecologists' definition is that a decline in environmental quality has an adverse impact on the welfare of the society. Therefore development can only be sustained by maintaining (or increasing) net wealth, environmental quality, and the stock of renewable resources.

11.2.3 The Sociologists' Perspective

From the sociologists' viewpoint, the key actors on the economic-environmental system are human beings. Human beings' patterns of organisation are important for developing viable solutions to SD. From this perspective, failure to devote attention to social factors in the development process will seriously hamper the progress of programs and policies aimed at achieving SD.

11.2.4 The Intergenerational and Intragenerational Equity Perspectives

The intergenerational equity perspective is more restrictive than the previous definitions. It suggests that the rate at which natural resources are being exploited is too fast and works against the interests of the unborn. As such, sustainable development is defined as the maintenance of non-declining per capita income over the indefinite future (the neoclassical definition) and maintaining the stock of renewable resources (ecological definition) with the added proviso that the net value of the stock of non-renewable resources must also be non-declining. According to this view, if the above conditions are met, our quality of life will not only be maintained but also future generations will have an undiminished or even enhanced stock of natural resources and other assets. It is also consistent with intra-generational equity since it is argued that non-declining natural capital implies that per capita utility or well-being is also being maintained.

11.2.5 The Materials Balance Approach

The materials balance approach to SD is derived from the First and Second Laws of Thermodynamics, which recognise physical and ecological constraints to economic activity (Chapter 2). The First Law of Thermodynamics states that energy cannot be created or destroyed. This implies that, although the form of energy may change, the total amount of energy in the system remains constant. The Second Law of Thermodynamics states that entropy always increases. A major implication of the First Law for

SD is that any raw material inputs used in the production and consumption process must eventually be returned to the environment as high-entropy waste products or pollutants. Recycling can help to reduce the amount of waste, to some extent, although recycling cannot be 100 percent effective. The Second Law implies that economic processes (i.e., production and consumption) are time irreversible in the sense that waste material cannot be fully converted to useful energy.

The materials balance approach basically raises doubt about human kind's ability to indefinitely extract more energy and materials from the world's ecosystem. The materials balance approach therefore disputes the neoclassical assumption that income growth leads to increase in human satisfaction. In this respect, the concept of the materials balance approach is similar to the ecologists' concept of SD.

11.2.6 *The Biogeophysical Perspective*

Sustainable development has also been defined on a biophysical dimension. According to Mohan Munasinghe and Walter Shearer, biogeophysical sustainability is the maintenance and/or improvement of the integrity of the life-support system on Earth. Sustaining the biosphere with adequate provisions for maximizing future options includes providing for human economic and social improvement for current and future human generations within a framework of cultural diversity while: (a) making adequate provisions for the maintenance of biological diversity and (b) maintaining the biogeochemical integrity of the biosphere by conservation and proper use of its air, water and land resources (Munasinghe and Shearer, 1995).

11.2.7 *The IUCN-World Conservation Union Definition*

The International Union of Conservation of Nature and Natural Resources (IUCN) defines SD as the maintenance of essential ecological processes and life support systems, the preservation of genetic diversity, and the sustainable utilization of species and ecosystems (IUCN, 1980). Their approach has an anthropocentric focus because it places emphasis on achieving a quality of life (or standard of living) that can be maintained for many generations. Sustainability is described in terms of fulfilling people's cultural, material, and spiritual needs in equitable ways.

11.2.8 The World Commission on Environment and Development's Definition

It appears that in defining sustainable development, the World Commission on Environment and Development (WCED) took into consideration all the above perspectives, in recognition of the fact that SD is a multifaceted concept. The WCED (or the Brundtland Commission) defined sustainable development as 'development that meets the needs of the present without compromising the ability of future generations to meet their own needs' (WCED, 1987:43).

11.2.9 Other Definitions

Various writers on the subject have put forth variants of the WCED definition. Pearce *et al.* (1989) have offered the following interpretation of SD:

> *Sustainable development involves a substantially increased emphasis on the value of natural, built and cultural environments...sustainable development places emphasis on providing the needs of the least advantaged in society ('intragenerational equity'), and on the fair treatment of future generations ('intergenerational equity')* (Pearce *et al.*, 1989).

R. Kerry Turner also described SD as follows:

> *Sustainable development involves maximising the net benefits of economic development subject to maintaining the services and quality of natural resources over time* (Turner, 1988).

Maurice Strong, defined SD as:

> *Sustainable development involves a process of deep and profound change in the political, social, economic, institutional, and technological order, including redefinition of relations between developing and more developed countries* (Strong, 1992).

Robert Solow has described 'sustainability' in quite different terms, referring to it as the 'duty' of humanity to future generations. According to him,

> *The duty imposed by sustainability is to bequeath to posterity not any particular thing...but rather to endow them with whatever it takes to achieve a standard of living at least as good as our own and to look after the next generation similarly* (Solow, 1992).

Similar definitions of sustainability in terms of intragenerational equity have been expressed by Toman (1992) and Redclift (1993). This conception of 'justice' or 'fairness' dates back to Aristotle who stated that justice of exchange has to be orientated towards equality. That is, one may not give less than what one has received. In that sense, an economy can be described as 'sustainable' if it allows future generations to have no less than what the current generation receives.

From the foregoing discussion, it can be seen that the pure neoclassical definition of SD can be rejected because 'development' or welfare involves more than mere increases in income (e.g., Gross National Product) over time. However, the role of economic growth in achieving SD must not be discounted. The WCED emphasised 'the essential needs of the world's poor, to which overriding priority should be given'. This statement implies that the aim of SD should be to enhance ordinary people's living standards, with particular attention to the less fortunate in society. The catch, however, is that to improve the lot of the poor, the economy needs to grow in order to generate the funds needed for poverty alleviation. In fact, the WCED itself projected that, in order for SD to take place, the economies of the developing world must grow at rates in the order of at least 3 percent per annum. Thus, we may conclude that economic growth is a necessary but not sufficient condition for sustainable development.

Today, many developing countries are caught in a vicious circle. In order to develop, they must grow. However, for most countries such growth entails exploiting their scarce natural resources, which, in turn, may pose risks to the environment. Many developing countries are saddled with huge external debts and many have chosen to use their natural resources to improve per capita incomes or to pay off their debts.

11.3   Conditions for Sustainable Development

From the foregoing discussion, it is important, for purposes of implementing strategies towards SD, to further investigate the conditions that are necessary to promote SD. Pearce *et al.* (1994) have proposed two conditions for sustainable development: **weak sustainability** and **strong sustainability**.

Weak sustainability assumes that there can be perfect substitution between physical, natural and human capital. For example, the depletion of natural capital in the current period can be offset by increases in physical or human capital (the neoclassical economics definition). Sustainable development can be achieved as long as the stocks of capital available to future generations are at least equal to the stocks available to the current generation. That is, so long as there is a constant aggregate capital stock. Strong sustainability, on the other hand, is much more restrictive. It assumes that not all natural capital stock can be substituted for man-made capital stock. For example, some ecosystem functions that are essential for the maintenance of living creatures cannot be replicated by human beings and the ozone layer cannot be recreated. Strong sustainability also implies that we have to preserve environmental assets for the future because of our limited understanding of their life-supporting functions.

The concept of strong sustainability is implicit in both the Brundtland Report and the *World Conservation Strategy*. According to the Brundtland Report, 'if needs are to be met on a sustainable basis the Earth's natural resource base must be conserved and enhanced' (WCED, 1987:57). The *World Conservation Strategy* makes references to maintaining 'essential ecological processes and life support systems' and 'sustainable utilisation of species and ecosystems' (IUCN, 1980:1).

Georgescu-Roegen (1971, 1979) has proposed the concept of **resilience** as a condition for SD. This requires the maintenance of a system within its thresholds of healthy and productive operation. The resilience perspective is based on an interdisciplinary understanding of the operation of ecological, environmental and social systems.

11.3.1 A Working Definition for Sustainable Development

From the foregoing discussion, it can be seen that the pure neoclassical definition of SD can be rejected because 'development' does not necessarily equate to 'economic growth (i.e. increases in per capita income). However, the role of economic growth in achieving SD must not be discounted. It has been estimated that the rate of economic growth necessary to keep the numbers of the poor from increasing is between 3–5 percent per annum. That is, in order to arrest the increase in poverty, the economy must grow at rates in excess of population growth rates. Economic growth is required to generate income to pay for the development process. A recurring theme in the above definitions is that SD involves improving the quality of life and maintaining environmentally responsible policies and practices. Improving the quality of life involves reducing poverty (which indirectly relieves pressures on the environment), improving health and nutritional status, improving education, and improving equity. To improve equity, policies must be developed to improve access to resources and achieve a 'fairer' distribution of income. To conserve environmental resources, there is a need to manage our natural resources wisely and to lower the intensity of resource use in order to leave future generations an unlimited (or even enhanced) stock of natural resources and other assets. A crucial element in the drive to improve the quality of life and the management of our natural resources is the reform of our institutions.[73] The role of institutions in achieving SD is discussed further below.

11.4   Measuring Sustainable Development

In this section we consider the issues of how to measure SD and translate it into action. We begin with the traditional measures of welfare or well-being, GDP and GNP. After a critique of these measures, we proceed to consider alternative measures of welfare.

11.4.1 Traditional Measures of Welfare

The current system of compiling national accounts, the UN's System of National Accounts (SNA) uses GDP and GNP as measures of welfare. Gross Domestic Product is the value of final goods and services produced in the domestic economy, whereas GNP is the value of goods and services produced in the domestic economy plus overseas income other than for exports. For aid purposes, the World Bank has divided countries into low, middle and upper income using income cut-offs based on these measures.[74] Measures such as GDP and GNP may provide a biased or overstated view of human welfare because they do not consider depreciation of the natural

---

[73] The 2003 World Development Report defines institutions as the rules and organisations, including informal norms that coordinate human behaviour.

[74] The per capita GNP cut-offs are: high income (US$9,266), middle income (US$756–9,265), and low income (US$755 or less).

capital stock as a result of economic exploitation and environmental degradation (see Box 11.2). GDP is calculated on the basis of the value of goods and services produced in the economy, and thus it will indicate an increase (i.e., an improvement in 'welfare') even when our stock of natural capital is being depleted and we have to expend more effort to extract lower grade resources. One anomalous property of GDP (or GNP) is that it will increase even when the quality of the environment is reduced by pollution. This is because it ignores degradation of environmental quality and effects on human health and welfare. Thus, for example, increase in the consumption of potentially health-threatening goods such as alcohol, cigarettes and fatty foods will appear as an increase in GDP, although they may even decrease individuals' welfare and impose a heavy burden on the national health budget. GNP increases when expenditures are made on pollution abatement. This is another distortion because environmental protection expenditures may actually be social costs of maintaining environmental quality (i.e. defensive expenditures). Other objections to the use of GDP (or GNP) as a measure of welfare include the following:

- It is an average measure that does not consider how income or wealth is distributed within the population. A country may have a high GDP per capita and be classified as 'rich' and yet there may be a high proportion of people who are deprived.
- It ignores changes in population and the productivity of human capital.
- It is a 'flow' concept (measures the flows of goods and services) and does not say anything about the stocks of natural resources. Thus it does not provide an adequate guide about how resources are being managed.
- It only accounts for transactions that occur in the market place and does not adequately consider subsistence production of peasants, housework (e.g., home cooking and gardening) and the work of voluntary or charitable organizations, and
- It does not discriminate between different types of goods. In resource dependent economies, GDP/GNP can provide an erroneous picture of welfare and economic progress (see Box 11.2). Therefore, any long term development plans based on such statistics may not be environmentally sound or sustainable.

Box 11.2 The Indonesian Case Study

David Repetto and others at the World Resources Institute in Washington, D.C., undertook a study to adjust the national accounts of Indonesia for environmental effects in 1990 (Repetto *et al.* 1989). Indonesia's natural resources consist of: oil, gas, minerals, timber and forest products. Together, these resources account for 44 percent of GDP, 84 percent of exports, and 55 percent of total employment.

Economic progress over the period of the study, 1970 to 1990, has been considered very good. For example, GDP per capita grew at 4.6 percent per annum between 1965 and 1986. This is quite high in relation to that of low-income and middle-income countries. Repetto *et al.* (1989) adjusted Indonesia's national accounts for for the period 1971–1983 by subtracting from GDP estimates of net national resource depreciation for only three sectors: petroleum, timber and soils, to arrive at an estimate of net domestic product (NDP).

The results clearly indicated that official GDP estimates overstated net income and growth of net income. The actual overstatement was higher since only three natural resources (petroleum, timber and soils) were considered. Items not included were: (i) exhaustible resources — natural gas, coal, tin, nickel and (ii) depreciation of renewable resources (e.g., fisheries and non-timber forest products).

Source: Repetto *et al.* (1989).

11.4.2 *Alternative Measures of Welfare*

Within the last decade alternative measures of welfare have been developed to rectify the deficiencies of GDP/GNP measures in order to provide a better guide of progress towards SD. From the definitions given in the previous section, it can be seen that it would be almost impossible to provide a single indicator that will capture the concept of SD. In this section, we will consider SD indicators under the broad categories of economic, ecological and sociopolitical. The economic indicators include **Environmentally Adjusted Domestic Product (Green Accounting)** and **Genuine Saving**. The sociopolitical indicators include the **Index of Sustainable Economic Welfare**, the **Genuine Progress Indicator**, and the **Human Development Index**. The ecological measures include **Net Primary Productivity** and

Carrying Capacity, Ecological Footprints, and **Environmental Space**. We begin the discussion with the economic indicators.

Economic Indicators: Environmentally Adjusted Product and Income (Green Accounting)

The environmentally adjusted net domestic product, referred to as 'Eco Domestic Product' (EDP) and national income (ENI) is a result of the United Nations Statistical Office (UNSO) effort's to develop a new framework for integrated environmental and economic accounting, referred to as the integrated System of Environmental and Economic Accounts (SEEA) (Bartelmus *et al.*, 1992). Details of the SEEA are given in next chapter. The EDP is obtained by adjusting the national accounts for changes in the quality of the natural environment and the depletion of natural resources.

Measures of EDP can be used to assess progress towards SD in the following ways:

- It can be used to indicate whether economic growth is sustainable and whether there are structural distortions in the economy by following environmentally unsound production and consumption patterns.

- For example, let C = total consumption expenditure; then C/EDP computed from the SEEA could be used as an indicator to indicate non-sustainable growth patterns. If C/EDP > 1, then it can be implied that the country is running down its natural capital base and therefore growth is not sustainable. On the other hand, C/EDP < 1 implies that the capital stock is being left intact or enhanced.

- A high rate of depletion of natural resources would appear in the accounts as a low rate of capital accumulation, indicating environmentally unsound production and consumption patterns.

Economic Indicators: Genuine (Extended) Saving

The concept of Genuine (or Extended) Saving (GS) is a new measure of sustainability. A country's wealth can be defined in terms of the value of a whole range of assets including the value of physical, natural and human capital. Genuine Saving measures the change in the total value of the three components of resources in the economy over a given period of time, usually a year. GS helps to track the sustainability of the economy by incorporating the effects of natural capital depletion into the standard flow of savings and income. For example, GS falls when the extraction or depletion of natural resources is not balanced by an offsetting re-investment in natural, physical or human capital. Further discussion of GS can be found in the next chapter.

Sociopolitical Indicators: Index of Sustainable Economic Welfare

The index of sustainable economic welfare (ISEW) was proposed by Daly and Cobb (1989) as an alternative measure of welfare changes and a better indication of economic progress than GDP/GNP. The computation of ISEW normally begins with compuation of personal consumption expenditures which are weighted with an index of income inequality. Then, certain welfare-relevant contributions such as services of hosuehold labour and the services of streets and highways are added. On the other hand, certain welfare-relevant losses such as 'defensive expenditures', costs of environmental pollution, costs of depletion of non-renewable resources, and long-term environmental damage are subtracted from the estimates. The ISEW can be expressed in equation form as

$$ISEW = C + P + G + W - D - E - N \qquad (11.1)$$

where: C = weighted consumer expenditure; P = non-defensive public expenditures; G = growth in capital and net change in international position; W = estimate of non-monetarised contributions to welfare; D = defensive expenditures; E = costs of environmental degradation; and N = depreciation of natural capital.

The ISEW is expressed in monetary units and a rising value would imply that the economy is becoming more sustainable. On the other hand, a falling value would indicate an unsustainable path.

Sociopolitical Indicators: Genuine Progress Indicator

The Genuine Progress Indicator (GPI) was proposed by Cobb *et al.* (1995). The GPI is an extension of the ISEW and aims to provide an indicator that more accurately reflects the health of the economy. The difference between the GPI and the ISEW is that the GPI excludes both public and private defensive expenditures on health and education, includes deductions of cost estimates for loss of leisure time, unemployment, but excludes loss of forests. As with the ISEW, a rising GPI indicates that an economy is heading towards a sustainable path, while a falling GPI indicates the opposite.

The major criticism of both the GPI and ISEW is that they involve a series of *ad hoc* adjustments. Also, like the EDP, they are based on current flows rather than stocks and thus they do not really address the maintenance of capacity, which, it has been argued, is at the heart of the concept of sustainability.

Sociopolitical Indicators: Human Development Index
Human development is about much more than the rise or fall of national incomes. Thus, relying on GDP/GNP will give a narrow view of economic development because it focuses only on income. The UN has developed an alternative measure, the HDI, which measures a country's achievements in three aspects of human development: longevity, knowledge, and a decent standard of living. Longevity is measured by life expectancy at birth; knowledge is measured by a combination of the adult literacy rate and the combined gross primary, secondary, and tertiary enrolment ratio; and standard of living is measured by GDP per capita (PPP US$). As a rough index of development, non-declining HDI over time would indicate a sustainable human development path. However, because the HDI considers only the educational and health status of the population, it too provides a limited view of development.

Ecological Indicators: Net Primary Productivity and Carrying Capacity
Net Primary Productivity (NPP) (Vitousek *et al.*, 1986) is a measure of the total available food resource for a system. NPP is derived from the ecological notion of carrying capacity (K), which is defined as the maximum population size that a given area can support without diminishing its ability to support the species in future periods. Vitousek *et al.* (1986) have linked NPP to K at the global level and have concluded that world NPP would not be able to support predicted increases in world per capita consumption rates. The NPP/K ratio for any given country can thus be interpreted as indicating how close or how far from its carrying capacity that country is. Thus, for example, NPP/K=1 indicates a sustainable population level, given current food requirements or food consumption.

The main criticism of the NPP/K indicator of SD is that it only considers the relationship between consumption and biological production. However, at a sub-global level, consumption can exceed natural productivity through imports. It also ignores the fact that natural productivity can be enhanced through the use of man-made capital and non-renewable natural resource inputs.

Ecological Indicators: Ecological Footprints

The concept of ecological footprints was developed by Wackernagel and Rees (Rees and Wackernagel, 1994; Wackernagel and Rees, 1996). It is a land-based measure of SD, which compares human demands in a given country, in terms of consumption, with the extent to which those demands can be met from the land area in that country. Energy, food and timber consumption per capita are expressed in terms of the land areas needed to produce these amounts. This is similar to the NPP/K indicator discussed above. A given country's 'footprint' is then calculated in relation to the available land area (excluding unproductive land). In this approach, a positive footprint (also referred to as an 'ecological deficit') could mean that the country's natural capital is being depleted, or alternatively, it could mean that it is imposing part of its footprint on other countries via imports. The ecological footprint concept assumes that the only sustainable form of energy is that derived from renewable resources.

The main criticism of the ecological footprints measure is that it is not a predictive measure. That is, given the value of this year's measure, we are not able to predict whether next year's measure will be higher or lower. However, this deficiency is also true of other measures such as EDP, Genuine Saving, ISEW and GPI.

Ecological Indicators: Environmental Space

The concept of Environmental Space is concerned with the fairness of resource use in any given country relative to world average use of that resource. The environmental space approach involves comparing global average use of a given resource (in per capita terms) with national consumption (also in per capita terms). Selected resources often include non-renewable resources, arable land, forestry, and water resources. Pollutants are included in the measure by specifying an upper limit on assimilative capacity. For example, in a study involving CO_2, Friends of the Earth (1995) found average per capita emissions in Europe to be 7.3 tonnes compared to a predicted per capita maximum of 1.7 tonnes per annum, allowing for an estimated world population of about 7.19 billion people by 2010.

11.5 Operationalising the Concept of Sustainable Development

Although there are many definitions of SD, practical measures to translate them into specific plans are in short supply. What is required is a suitable framework that will enable plans and policies to be tested to see whether they are economically, ecologically, and socially sustainable prior to implementation. One such approach is the environmentally sustainable development (ESD) triangular framework (Serageldin and Steer, 1994). The triangular approach may be explained with the aid of Figure 11.1. The figure shows the links between the environment, the economy and the society and draws attention to the need to strike a balance between the three components in order to achieve SD. Thus, for example, a proposal first has to be economically and financially sustainable in terms of enhancing growth and making efficient use of scarce resources. Secondly, it must be ecologically sustainable. This can be determined in terms of its impacts in terms of ecosystem integrity, carrying capacity and conservation of natural resources including biodiversity. Finally, it must be sustainable in terms of certain social criteria. These include equity, social mobility, social cohesion, participation, empowerment, cultural identity, and institutional development. As has been shown, the social dimension is important for achieving SD. Neglect of the social dimension will lead to a collapse of institutions, increase social disorder and negatively impact on the environment.

How can we ensure that the triple objectives of economic, ecological and social integrity of plans, programs and policies are met? This could be done in a number of steps. The first step would be to develop a system of environmental and social indicators that could be used in decision-making processes. This issue has been briefly discussed above and is dealt with in more detail in the following chapter (Green Accounting and the Measurement of Genuine Saving). The next step would be to develop new ways, or modify existing approaches, for incorporating this information into economic decision-making processes. In this section, we consider modifications to cost-benefit analysis to incorporate environmental effects. The third step would be to restructure institutions to provide the correct signals for sustainable resource use and to achieve distributional goals. This could be achieved by implementing market-based incentive mechanisms correcting for any incidence of policy failure.

Figure 11.1 The Environmentally Sustainable Development Triangle

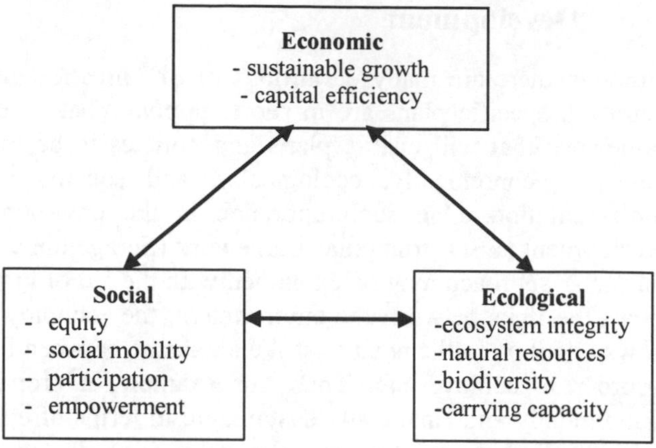

```
                    ┌──────────────────────────┐
                    │        Economic          │
                    │    - sustainable growth   │
                    │    - capital efficiency   │
                    └──────────────────────────┘
                         ↗              ↘
   ┌────────────────────┐              ┌──────────────────────┐
   │       Social       │              │      Ecological      │
   │   - equity         │ ←─────────→  │  -ecosystem integrity│
   │   - social mobility│              │  -natural resources  │
   │   - participation  │              │  -biodiversity       │
   │   - empowerment    │              │  -carrying capacity  │
   └────────────────────┘              └──────────────────────┘
```

Source: adapted from Serageldin and Steer (1994:2).

11.5.1 *Establishing a New System of Economic Analysis*

The way in which government agencies appraise investment projects is crucial in efforts to achieve SD. Using the traditional cost-benefit technique, a project is potentially acceptable if benefits minus costs, suitably discounted, are greater than zero. Environmental costs are usually excluded from this analysis. There is a need to modify this procedure to ensure that costs include any environmental damage, and benefits include any environmental gains from the project.

A modification to the project selection criterion in social CBA is to select a project if:

$$\text{NPV} = \frac{\sum_{t=1}^{n}(B_t - C_t - E)}{(1+r)^t} > 0 \qquad (11.2)$$

where: E = net environmental benefits
           B = total commercial benefits
           C = total costs
           r = social discount rate

It is to be noted that E (net environmental benefits) is to be treated as an opportunity cost, i.e., forgone benefits, and subtracted from the commercial benefits.[75] Recall that in Chapter 5, we defined total economic value (*TEV*) as:

$$TEV = TUV + TNUV \qquad (11.3)$$

where: TUV = total use value
TNUV = total non-use or passive use value

Equation (11.2) can be substituted into Equation (11.1) and the decision rule becomes:

$$NPV = \frac{\sum_{t=1}^{n}(B_t - C_t - (TUV + TNUV)}{(1+r)^t} > 0 \qquad (11.4)$$

When applied to a portfolio of projects, Equation (11.1) may be regarded as a weak sustainability rule because it implies that the discounted present value of net benefits across all projects must be positive. However, there could also be a strong sustainability rule. Such a rule would say that net environmental costs across all projects must be positive in every time period.

The Safe Minimum Standard of Preservation

In many resource allocation decisions the choice is often between preserving a given natural environment or destroying it in order to undertake some form of development. The American economist, S.V. Ciriacy-Wantrup defined the **safe minimum standard of preservation** (SMS) as the minimum critical mass required for species preservation (Ciriacy-Wantrup, 1968; Bishop, 1978). In other words, the SMS is a level of conservation that is high enough to reduce the probability of extinction or irreversible loss of a species to a low level. This concept provides a decision rule for choosing between preservation and conservation. It is known as the 'Sustainability Approach', and can be stated as follows:

---

[75] In some cases E could be a benefit in which case it should be added to commercial benefits.

Avoid irreversible environmental damage unless the
social cost of doing so is unacceptably large.

Box 11.3 Evaluating the Gordon River Hydroelectric Proposal, Tasmania

H. Saddler and others from the Australian National University evaluated
two options for electricity development on the Gordon River in Tasmania
in 1980. The options were between coal and hydroelectric power. The
former was more expensive. However, the latter involved flooding the
Gordon River and destroying a vast wilderness area. Although they did not
estimate preservation benefits directly, they estimated that if preservation
benefits were to be A\$725,000 in the first year, then the present value of
preservation benefits will exceed the difference in costs between the two
options. The authors leave the judgement to the decision-maker. In the
end, the Gordon River dam was never built, due mainly to the activities of
environmentalists. However, it is clear that the annual preservation
benefits of the unique wilderness area exceed A\$725,000 per annum (even
without doing a contingent valuation survey) and therefore the correct
decision, using the SMS approach, is not to build the dam.

Source: Saddler (1980).

Asafu-Adjaye (1991b) outlines how the SMS concept could be used to
modify the CBA procedure so as to address concerns about sustainability
and intergenerational equity. The main problem with this approach is
deciding what is 'unacceptably' large. Obviously, this involves a value
judgement (Box 11.3). In some developing countries where there is abject
poverty and hunger, it may be the case that preserving the environment in its
natural state (i.e., the no development option) may impose social costs that
are unacceptable unless there are alternative means of income generation.

11.6   Constraints to the Implementation of Sustainable Development

The UN General Assembly Resolution 47/190 endorsed the concept of
sustainable development, as embodied in Agenda 21, and urged its
implementation at the national, regional and global levels. A Commission on

Sustainable Development (CSD) was created in 1992 to oversee the implementation of Agenda 21, and to monitor and report on implementation of the Earth Summit agreements at the local, national, regional and international levels. Implementation of Agenda 21 is the responsibility of the individual governments. Each government, if necessary with input from the aid agencies, is supposed to come up with strategies for sustainable development. To date, many countries have complied with this requirement. Furthermore, most of these countries have instituted coordination councils or committees, some of which are chaired by high-ranking government officials. For example, the Prime Minister of Australia chairs the Inter-Governmental Committee on Ecologically Sustainable Development. The Committee drafted a National Strategy for Ecologically Sustainable Development, a National Greenhouse Response Strategy and an Inter-Governmental Agreement on Environment.

To date, the main impediments to the successful implementation of SD, especially in developing countries, are lack of institutional capacity and inadequate financial resources. At a five-year review of Earth Summit progress in 1997, the CSD reported that about 150 countries had either established new or modified existing institutional structures to assist in the implementation of Agenda 21. However, in some developing countries, these structures are not well-developed. In cases where they have been established (e.g. National Council for Sustainable Development), their effectiveness has been compromised by problems such as lack of authority to enforce policies, insufficient representation, lack of qualified personnel and lack of financial resources. Some countries have instituted legislation to protect the environment. However, some of these regulations literally have no teeth due to lack of trained personnel to monitor and enforce them.

The Asian Development Bank (ADB) estimates that the cost of implementing some of the Agenda 21 initiatives in the Asia-Pacific region will amount to about US$13 billion by the year 2000, rising to US$70 billion by 2010 (ESCAP and ADB, 1995). It is quite clear from this that sustainable development does not come cheap. Many developing country governments are facing severe budget constraints that compromise their ability to implement SD programs. All over the world, official development assistance (ODA) as a proportion of GNP has declined. The challenge to developing country governments is to generate more private sector interest and involvement in SD.

11.7   Summary

This chapter began with an overview of definitions of sustainable development. Definitions were presented from 'neoclassical' economics, ecology, materials balance and intergenerational equity perspectives. Combining all these perspectives, sustainable development can be characterised by a vector of socially desirable objectives. To achieve SD this vector must be non-declining over time.

An important issue that was discussed was the conditions for achieving SD. It was argued that although economic growth is a necessary requirement for the attainment of SD, it is not sufficient to guarantee SD. Two sufficient conditions for SD—weak and strong sustainability—were introduced. Weak sustainability assumes perfect substitutability between different forms of capital and therefore requires a constant aggregate capital stock (i.e., physical, natural and human capital) to be maintained. On the other hand, strong sustainability recognises that some forms of capital are not easily substitutable and thus requires a constant stock of all natural capital to be maintained.

It was argued that the current measures of welfare or economic progress are inadequate because, among other things, they ignore the impacts of environmental degradation and changes in the quantity and the quality of the natural capital stock. Alternative measures of welfare were presented and their limitations were discussed.

We discussed how the concept of SD could be operationalised through the policy process. Policy measures are required to ensure that prices reflect environmental costs. Furthermore, there is a need for institutional restructuring to achieve distributional goals. Modifications to the current system of economic analysis were suggested in order to incorporate sustainability. The main constraints to implementation of SD are lack of financial resources and lack of institutional capacity. Transfer of EFTs is urgently required to speed up progress towards achievement of the Agenda 21 objectives.

Key Terms and Concepts

ecological footprints   index of sustainable economic
environmental space   welfare

environmentally adjusted domestic product
environmentally adjusted national income
ESD triangular framework
genuine (or extended) saving
genuine progress indicator
green accounting
gross domestic product
gross national product
human development index

net primary productivity and carrying capacity
resilience
safe minimum standard of preservation
strong sustainability
sustainable development
system of environmental and economic accounts
system of national accounts
weak sustainability

Review Questions

1. List the main features of sustainable development as defined by neoclassical economists and ecologists.

2. State the weaknesses in the economist's view of sustainable development.

3. Explain the difference between the weak and strong sustainability conditions.

4. List the shortcomings of the current System of National Accounting.

5. What is the Safe Minimum Standard of Preservation?

6. List some of the problems that countries face in trying to implement programs for sustainable development.

Exercises

1. Suppose biologists have come up with an estimate of the cost of ensuring the survival of an endangered species (e.g., the Sumatra tiger). Policy makers are faced with the choice of protecting this animal's

habitat. Using what you have learned from this and previous chapters, discuss what kinds of information needs to be collected and how it should be used. Give particular attention to how benefits can be determined and what kind of economic analysis can be performed.

2. Some have argued that given our lack of knowledge about our ecosystem and biological diversity, we should exercise caution and be risk averse when it comes to exploiting environmental resources. Others have argued that for some countries, especially the poorer ones, such a policy could even be risky because they would be deprived of the necessary capital to improve their circumstances. Discuss both sides of this debate.

3. It has been argued that every unit of a non-renewable natural resource extracted today precludes its use by future generations. Do we have a moral responsibility to use only renewable resources? Do you agree or disagree? State your reasons.

4. The traditional measures of welfare (GDP and GNP) have been criticised on various grounds. The Index of Sustainable Economic Welfare, the Human Development Index, Net Primary Productivity and Carrying Capacity, and Ecological Footprints have been proposed as alternative measures of welfare. Explain each of these indices and discuss their limitations as measures of sustainable development.

References

Asafu-Adjaye, J. (1991a). Environmental Accounting in Papua New Guinea. *Pacific Economic Bulletin*, 8: 42-49.

Asafu-Adjaye, J. (1991b). Sustainable Economic Development: Implications for the Analysis of Development Projects in LDCs. *International Journal of Development Planning Literature*, 6: 57-66.

Bishop, R. (1978). Endangered Species and Uncertainty: The Economics of a Safe Minimum Standard. *American Journal of Agricultural Economics*, February, 10-18.

Bartelmus, P.L.P. (1992). Environmental Accounting and Statistics. *Natural Resource Forum*, February: 77-84.

Ciriacy-Wantrup, S.V. (1968). *Resource Conservation: Economics and Policies.* 3rd Edition, University of California, Division of Agricultural Science, Berkeley and Los Angeles.

Cobb, C., Halstead, E., and Rowe, J. (1995). *The Genuine Progress Indicator – Summary of Data and Methodology*, Redefining Progress, Washington, D.C.

Costanza, R. (1994). Environmental Performance Indicators, Environmental Space and the Preservation of Ecosystem Health. In *Global Change and Sustainable Development in Europe,* Manuscript on file at the Wuppertal Institute, Nordrhein-Westfalen, Germany.

Daly, H. and Cobb, C. (1989). *For the Common Good*, Beacon Press, Boston.

Daly, H.E. (1991). *Steady-State Economics*. Island Press, Washington.

Economic Commission for Asia and the Pacific (ESCAP) and Asian Development Bank (ADB) (1995). *State of the Environment in Asia and the Pacific*. United Nations, New York.

Friends of the Earth Europe (1995). *Towards a Sustainable Europe*, Friends of the Earth, Brussels.

Georgescu-Roegen, N. (1971). *The Entropy Law and the Economic Process*, Harvard University Press, Cambridge, Mass.

Georgescu-Roegen, N. (1979) 'Energy analysis and economic valuation' in *Southern Economic Journal* 45(4): 1023-58.

Government of Australia (1992). *National Strategy for Ecologically Sustainable Development,* Australian Government Publishing Service, Canberra.

International Union for the Conservation of Nature and Natural Resources, IUCN. (1980). *World Conservation Strategy*. Gland, Switzerland.

Jorgensen, D.W. and P.J. Wilcoxen (1990). Environmental Regulation and US Economic Growth. *Rand Journal of Economics*, 21: 314-40.

Munasinghe, M. and Shearer, W. (1995). An Introduction to the Definition and Measurement of Biogeophysical Sustainability. In *Defining and Measuring*

Sustainability: The Biogeophysical Foundations, M. Munasinghe and W. Shearer, ed. Washington D.C. Distributed for the United Nations University by the World Bank.

Pearce, D. W., K. Hamilton, et al. (1996). Measuring sustainable development: progress on indicators. *Environment and Development Economics*, 1(1): 85-101.

Pearce, D., Markandya, A. and Barbier, E.B. (1989). *Blueprint for a Green Economy*. Earthscan, London.

Pezzey, J. (1989a). Definitions of Sustainability. CEED Discussion Paper No. 9, Centre for Economic and Environmental Development, London.

Pezzey, J. (1989b). Sustainability, Intergenerational Equity, and Environmental Policy. Discussion Paper in Economics, No. 89-6, University of Colorado, Department of Economics, Boulder, Colorado.

Redclift, M. (1993). Sustainable Development: Needs, Values, Rights. *Environmental Values*, 1: 3-20.

Rees, W.E. and Wackernagel, M. (1994). Ecological Footprints and Appropriate Carrying Capacity: Measuring the Natural Requirements of the Human Economy. In *Investing in Natural Capital: The Ecological Economics Approach to Sustainability*, W.E. Rees and M. Wackernagel, eds., Island Press, Washington, D.C.

Repetto, R., Magrath, W., Wells, M., Beer, C., and Rossini, F. (1989). *Wasting Assets, Natural Resources in the National Income Accounts*. World Resources Institute, Washington, D.C.

Saddler, H. (1980). *Public Choice in Tasmania: Aspects of the Lower Gordon River Hydroelectric Development Proposal*. Centre for Resource and Environmental Studies, Australian National University, Canberra.

Serageldin, I. and Steer, A., eds., (1994). *Making Development Sustainable: From Concepts to Action*, Washington, D.C. : World Bank.

Solomon, A. (1990). *Towards Ecological Sustainability in Europe: Climate, Water Resources, Soils and Biota*, IISA RR-90-6, Laxenburg, Austria. 1990.

Solow, R. (1956). A Contribution to the Theory of Economic Growth. *Quarterly Journal of Economics,* 70: 312-20.

Solow, R.M. (1992). An Almost Practical Step Toward Sustainability. Invited Lecture, 40th Anniversary of Resources for the Future, Washington, D.C., October.

Strong, M. (1992). Required Global Changes: Close Linkages Between Environment and Development. In *Change: Threat or Opportunity*. Uner Kirdar, ed. NY: United Nations.

Tolba, M. (1987). *Sustainable Development—Constraints and Opportunities*, Butterworth, London.

Toman, M.A. (1992). The Difficulty in Defining Sustainability. *Resources for the Future*, Winter, No. 106.

Turner, R.K. (1988). Pluralism in Environmental Economics: A Survey of the Sustainable Economic Development Debate. *Journal of Agricultural Economics*, 39: 353-359.

United Nations (1992). *Agenda 21: Program of Action for Sustainable Development*. Final text of agreements negotiated by Governments at the United Nations Conference on Environment and Development, 3-14 June, Rio de Janeiro, Brazil.

United Nations Conference on the Human Environment, UNCHE (1972). *Development and Environment*. United Nations and Ecole Practique des Hautes Etudes, Mouton.

United Nations Development Program, UNDP (1998). *Human Development Report*. Oxford University Press, New York and Oxford.

Vitousek, P., Ehrlich, P., Ehrlich, A., and Matson, P. (1986). Human Appropriation of the Products of Photosynthesis. *Biosccience*, 36:368-373.

Wackernagel, M. and Rees, W. (1996). *Our Ecological Footprints: Reducing the Impact on the Earth*, New Society Publishing, Gabriola Island, BC.

World Commission on Environment and Development, WCED (1987). *Our Common Future*. Oxford University Press, New York and Oxford.

12. Green Accounting and Measurement of Genuine Saving

Objectives

After studying this chapter you should be in a position to:

❑ explain the basic characteristics of green accounting;

❑ explain the limitations of green accounting;

❑ explain the concept of genuine (extended) saving; and

❑ explain the limitations of measures of genuine saving

12.1 Introduction

The environment plays a number of important roles, including sustaining basic life support systems, providing raw material inputs to producers and consumers, serving as a receptacle for the waste products of producers and consumers, and providing amenities to consumers (e.g., recreation). In the periods when the world's population and the scale of economic activities were relatively small, environmental inputs were often regarded as 'free' goods and the environment was treated as a 'sink' for disposal of waste. However, there is a limit to the extent of the environment's capacity to assimilate waste. Pollution and environmental degradation begin to occur when this assimilative capacity is reached. Furthermore, once this limit is exceeded the ability of the environment to provide other services (e.g., provide inputs) is compromised. There is a need to view the natural environment not only as a resource but also as an asset similar to traditional assets such as land, labour and capital. The value of this resource must therefore be integrated into the economic system.

In Chapter 11, the deficiencies in the traditional System of National Accounts which places emphasis on GDP/GNP measures were discussed. Measures such as Net Domestic Product (NDP), while better than GDP for measuring sustainability, account only for the depreciation of produced assets, and ignore the depreciation of natural resources and degradation of the environment. Alternative 'greener' measures such as Green Accounting and Genuine Saving were briefly introduced in that chapter. This Chapter discusses the two approaches in a little more detail, noting their limitations. The chapter concludes with case studies reporting applications of the two approaches.

12.2 Green Accounting

Ahmed *et al.* (1989), Repetto *et al.* (1989) and Hartwick (1990). The approach taken by Hartwick is based on neoclassical growth models and attempts to specify 'optimal' adjustments to the SNA. The second approach, that of Ahmed et al. (1989) and Repetto et al. (1989), is a more practical approach that attempts to make piecemeal changes to the widely accepted SNA.[76] As of now, there is no universal agreement among economists as to how these adjustments should be made to reflect environmental damage. In December 1993, the UNSO, in collaboration with international agencies, launched the SEEA, which is based on the framework presented by Ahmed et al. (1989)[77]. The SEEA is implemented in the form of satellite (or supplementary) accounts that are linked with the core accounts of the SNA. In that sense, the SEEA maintains concepts and principles embodied in the SNA. The basic features of the SEEA are as follows:

- **Segregation and elaboration of all environment-related flows and stocks of traditional accounts.** This aspect of the SEEA seeks to identify the part of GDP which reflects the costs necessary to compensate for the negative impacts of economic growth, i.e., defensive expenditures.

---

[76] Excellent reviews of this literature can be found in Hamilton (1994) and Hanley (1997).

[77] The five agencies are the Commission of the European Communities, IMF, OECD, UN and World Bank.

- **Linkage of physical resource accounts with monetary environmental accounts and balance sheets.** This component attempts to establish comprehensive physical resource accounts to be linked to the monetary balance sheet and flow accounts of the system of national accounts. The resource accounts will consider the total reserves of natural resources and changes therein even when such resources are not yet affected by the economic system.

- **Assessment of environmental costs and benefits.** This key part of the SEEA seeks to improve on the SNA by accounting for depletion of natural resources and changes in environmental quality due to economic activity.

- **Accounting for the maintenance of tangible wealth.** In this component, natural capital is handled in the same way as physical capital. Natural capital, as defined here, includes renewable resources (e.g., forestry, fisheries,), non-renewable resources (e.g., land, soil, mineral), and air and water resources.

- **Elaboration and measurement of indicators of environmentally adjusted product and income.** The intention here is to develop modified macroeconomic measures of national income such as environmentally adjusted domestic product (EDP). This measure accounts for the costs of depletion of natural capital and changes in environmental quality.

As indicated earlier, EDP is obtained by subtracting from NDP the costs of environmental degradation and the depreciation of the stock of natural resources. Two alternative measures of EDP have been suggested:

EDP1: NDP minus depreciation of natural resources due to extraction; and
EDP2: EDP1 minus the cost of environmental degradation.

The construction of the SEEA proceeds in four of steps (see Figure 12.1). The first step involves compilation of physical accounts. These are nonmonetary accounts that measure resource depletion and environmental effects.

Figure 12.1 Framework of the system of integrated economic and environmental accounting

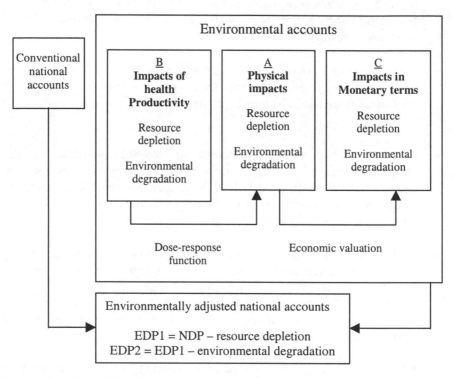

Source: adapted from Serageldin and Steer (1994).

12.2.1 Issues and Limitations of Green Accounting

Although progress has been made in terms of trying to change the way national accounts are compiled with the view to making them reflect environmental considerations, several practical difficulties and issues remain to be resolved. These issues include: accounting for defensive expenditures and pollution damages, the valuation of environmental damages, the treatment of transboundary pollutants, and the treatment of sociopolitical development goals and issues.

Defensive expenditures and pollution damages

'Defensive expenditures' can be defined as expenditures incurred by households and governments to reduce the effects of pollution. Examples of defensive expenditures for the household include buying water purification equipment to improve drinking water quality or buying a malaria prophylactic. For the government, this could include expenditures on litter removal or repairing degraded recreational sites. There is no agreement on how to handle these expenditures. Maler (1991) argues that such expenditures should not be deducted from NDP if the changes in the values of 'environmental services' (e.g. air and water quality) are included, since this would amount to double counting. Dasgupta (1995), on the other hand, states that defensive expenditures should be included in final demand. Bartelmus and van Tongeren (1994) argue that the cost of restoring polluted or damaged natural environments to their original state at the beginning of the accounting period should be deducted from NDP. This is similar to the EDP2 measure mentioned above and is referred to as the 'maintenance cost' approach to accounting for environmental effects. However, this maintenance cost approach has also been criticised on two grounds. First, Aaheim and Nyborg (1995) argue that costs that would have been incurred to prevent emissions from being higher this period than the previous period could severely underestimate the cost of actually repairing the damage, especially if the damage is irreversible. Second, the maintenance cost approach is problematic for cumulative pollutants for which any emission adds to the stock because there is no assimilative capacity. Hueting et al. (1992) argue that a better approach would be to value environmental damage at the cost of bringing environmental quality to 'sustainable levels'. However, the problem with this suggestion is that there is likely to be disagreement over how to determine what level is sustainable.

Valuation of environmental damages

Regardless of which measure of environmental damage to use, there are difficulties in placing monetary values on effects that are non-market in nature. As was discussed in Chapter 5, techniques such as contingent valuation, choice experiments, hedonic pricing and travel cost methods have been developed to assist environmental valuation. However, there are some environmental effects that are difficult to value even with the state of the art techniques. Even when such effects can be valued, there is the additional issue of whether the society's valuation is equal to the sum of the individual valuations. A practical consideration is that the different techniques often yield different estimates, implying that there could be large differences in the value of environmental depreciation used to compute EDP.

Transboundary pollutants
The transboundary issue concerns the case where some pollutants (e.g. SO_2 and NO_2) are emitted in one country and have adverse effects on other countries. An extreme case is where the effects (e.g global warming) are worldwide. The practical issue here is how, or whether, we should account for the polluting effects esternal to the country for which EDP is being computed. The same question applies to the case of imports of transboundary pollutants. Some people (e.g. Maler, 1995) have argued that pollution damages caused by a given country on other countries should be deducted from its NDP, since NDP measures the welfare impacts of projects in that country. However, he suggests that imports of emission into the country should be ignored. His suggestion raises a couple of issues. The first has to do with the practical problem of separating out pollution impacts in the country, which is made more difficult if there are also similar emissions from the country. The second point is the counter argument that if NDP is a measure of the welfare of the inhabitants of a given country, then imports of emissions should be counted but exports should not. Whatever the approach adopted, there is the additional problem that all countries should reach a consensus for the SEEA to be meaningful at the global level.

Treatment of sociopolitical development goals and issues
Development goals such as equity, cultural aspirations or political stability are difficult to quantify and quite impossible to value in monetary terms. Such effects would have to be specified in normative terms as targets or standards. In many developing countries today, there is increasing poverty, income inequality, corruption, and crime. These issues are difficult to measure, even when we use non-economic indicators.

12.3 Genuine (Extended) Saving

The concept of genuine (or extended) saving (Pearce and Atkinson, 1993) has been proposed as a broad measure of sustainability that values changes in the natural resource base and environmental quality in addition to man-made (or produced) assets. The traditional measure of a country's rate of wealth accumulation is gross saving which is given by the difference between GNP and consumption. Gross saving is the total amount that is set aside for the future in terms of either foreign lending or investment of productive assets, and it tells us little about whether or not a particular development path is sustainable. This is because we expect productive assets to depreciate. The issue then is whether the amount of depreciation is greater than gross saving. Net saving, which is gross saving minus depreciation of produced assets is a slightly better indicator of sustainability, although it focuses only on produced assets.

The World Bank has proposed a methodology for calculating Genuine Saving that they have applied to a number of countries. The process begins with the conventional national accounts. First, gross domestic saving is obtained by deducting net foreign borrowing (including net official transfers) from gross domestic investment, which consists of total investments in structures, machinery and equipment, and inventory accumulation. That is

$$Gross\ Domestic\ Saving = Gross\ Domestic\ Investment - Net$$
$$Foreign\ Borrowing + Net\ Official$$
$$Transfers \tag{12.1}$$

Next, net saving is obtained by deducting depreciation of produced assets from gross domestic saving. That is,

$$Net\ Saving = Gross\ Domestic\ Saving - Depreciation\ of$$
$$Produced\ Capital \tag{12.2}$$

Finally, GS is obtained by subtracting the value of resource depletion and pollution damages from net saving, and adding on the value of investment in human capital. That is,

$$GS = Net\ Saving + Human\ Capital\ Investment$$
$$- Depletion\ of\ Natural\ Resources - Pollution\ Damage \tag{12.3}$$

12.3.1 Human Capital Investment

The inclusion of human capital is based on the rationale that human capital contributes to overall sustainability by assisting in the ongoing creation and maintenance of national wealth. Human capital investment in the current period is assumed to provide the resources for improvements in productivity and income in the future. Although, historically, a number of specific measures of human capital have been developed (e.g. see OECD, 2000), in computing the GS measures, the World Bank uses current educational expenditures as a proxy for the value of human capital.[78] Obviously, this is not the best measure because expenditure by itself is an indicator of financial input into the process of human capital formation. Therefore, it is only an indirect indicator of the economic contribution of education to a nation's growth and productivity.

12.3.2 Depletion of Natural Resources

Depletion of natural resources is measured as the total rents on resources extracted and harvested. For non-renewable resources (e.g. bauxite, copper, gold, iron ore, and so on) rents are estimated as the difference between the value of production at world prices and the total production costs. For forest resources, the rent is computed as the difference between the rental value of log harvests and the corresponding value of natural growth and plantations less harvesting costs. In the computation of forestry rents, only the commercial value of timber is considered and other environmental services provided by trees such as carbon storage, watershed protection and the value of nontimber forest benefits are excluded. The calculation of resource rents also omits other natural assets such as fisheries resources and the economic costs of soil degradation.

12.3.3 Pollution Damages

Due to problems associated with precisely determining the welfare effects of pollution damage, a very simple approach is adopted in the computation of GS. Pollution damage is calculated only for CO_2, using a global estimate of US\$20 per metric ton of carbon emitted. It is, however, possible to extend the calculation to include other critical pollutants such as chloroflorocarbons (CFCs).

---

[78] Current educational expenditures consist of teachers' salaries and expenditure on textbooks.

12.3.4 Policy Uses of Genuine Savings Estimates

For developing countries that are dependent on natural resources for the bulk of their national income, GS estimates computed over a period of time can be used to answer pertinent questions about the sustainability of natural resource use. Such questions include:

- Are the rents from natural resources invested or consumed?
- What are the types of investments that a country makes?
- Do existing property rights regimes encourage sustainable exploitation of natural resources?
- Are royalties set appropriately to capture resource rents?
- Do policies to promote natural resource exports also contain plans for the investment of the resource royalties?
- What micro and macroeconomic policies such as government expenditures, taxation and interest rates can be devised to create incentives for higher genuine savings?

Some developing countries are beginning to experience pollution problems associated with rapid urbanisation. Computation of GS estimates would enable planners to answer the following questions

- Do the current pollution reduction policies target efficient levels of emissions?
- Are sufficient savings being made to offset cumulative effects of pollution?

12.3.5 Limitations of Genuine Saving Measures

There are a number of limitations associated with GS as a measure of sustainability. These can be divided into two categories – conceptual and technical. The conceptual issues deal with the way the measure is defined, while the latter deal with how the measure is actually computed. There are at least four conceptual problems with the definition of GS. First, it cannot be assumed that savings necessarily equates to investment. Furthermore, even if all savings are channeled into investment, this does not mean that output will be sustained. The sustainability of output will depend on the quality of the investment in capital and how efficient it is. For example, a developing country investing heavily in primary education may be setting itself up on a more sustainable growth path than one that is consuming the proceeds of natural resource extraction. Second, GS does not address the issue of intra-generational equity. For example, a country may have a high level of savings but these could belong to a small fraction of the population and therefore is not likely to lead to development. Third, GS is based on weak sustainability since it does not distinguish between the types of capital that can be substituted for each other. In that sense, it is a narrow concept of sustainability. Fourth, due to the fact that GS is measured in monetary units, changes in resource prices (which is usually beyond the control of individual countries) may cloud changes in physical stocks, and may therefore give a less than clear picture of changes in sustainability.

The main technical limitation of GS measures relates to the fact that it is currently based on the World Bank's methodology that excludes some important aspects of natural capital. For example, the approach restricts pollution damage to only CO_2, which is estimated at a flat rate of US$20 per ton for all periods. Other major air pollutants such as SO_x, NO_x, particulates, ozone and CFCs, as well as damage from water pollution, are excluded. The approach also restricts natural resource depletion to only two components—minerals and forestry—and makes no provision for the depletion of other land-based capital due to factors such as soil erosion, salinity and water pollution. Finally, the current approach also ignores the following important components of natural capital: freshwater and marine-based resources; the value of ecosystem services such as carbon sequestration, biodiversity and watershed regulation; the value of native remnant bushland (e.g. dry tropical savanna); the value of air resources; and industrial and household uses of water. Admittedly, most these omissions are due to the technical difficulties associated with estimating non-market values such as biodiversity and other ecosystem services. However, research advances should make it possible to at least obtain ball park estimates of some of these effects.

Case Study 12.1 A Calculation of Genuine Saving for Queensland[79]

Introduction

Estimates of GS were computed for the state of Queensland using the World Bank's methodology. Queensland is a natural resource dependent state with mining accounting for 45% of its natural capital, and agriculture and forestry accounting for 38% (See Figure 12.2). The GS for Queensland was estimated for the period 1989/90 to 1999/2000, at constant (1999/2000) prices. Some modifications were made to the calculation to reflect the State-level equivalent of each component in the World Bank's methodology. GS was defined as follows

$$GS = Net\ State\ Saving + Human\ Capital\ Investment - \\ Depletion\ of\ Natural\ Resources - Pollution\ Damage \qquad (12.4)$$

where

$$Net\ Saving = Gross\ Domestic\ Saving - \\ Consumption\ of\ Fixed\ Capital \qquad (12.5)$$

and

$$Gross\ Domestic\ Saving = Gross\ Domestic\ Investment \\ - Net\ Foreign\ Borrowing \\ + Net\ Official\ Transfers \qquad (12.6)$$

Gross Domestic Saving was measured using the State level equivalent of national disposable income, defined as national income plus net unrequited transfers. At the level of a State within a country, this implies adding the current account balance to Gross State Product (GSP). Human capital investment was measured by government expenditure on education. Effectively, this is a re-classification of government expenditure, as education expenditure is usually treated as an element of government consumption. It ignores other components of human capital such as health, and it measures the value of human capital in terms of the cost of education measured by public expenditure on the education sector.

---

[79] This section is based on Brown *et al.* (2003).

Following the World Bank's approach, pollution damage was restricted to include carbon dioxide only and water pollution damage was not included. Estimated carbon emissions were based on annual data for the whole of Australia and Queensland data for two years, 1989/90 and 1998/99. The intermediate years' values were estimated by interpolation using Queensland's share of total Australian emissions in the two end years. The calculation also restricted the calculation of natural capital depletion to forest and mineral resources. These two components represent the depletion of the economy's renewable and non-renewable resources, respectively, and makes no provision for the depletion of other land-based capital due to factors such as soil erosion, salination and water pollution. Furthermore, freshwater and marine-based resources are also excluded.

Figure 12.2 Components of natural capital for the state of Queensland

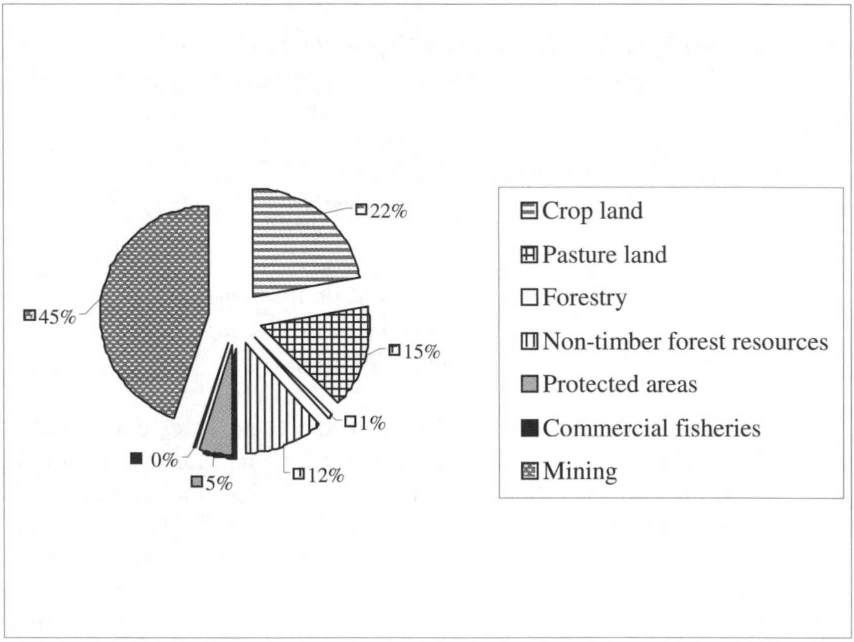

Resource rent for mineral production and forestry was estimated by first calculating gross operating surplus by adding royalties and company taxes to gross operating surplus. Next, 'normal profit' was calculated as 10% of total costs (including depreciation). This was then subtracted from gross operating surplus to arrive at the estimates of resource rents for mining and forestry.

Results

The results of GS estimates are graphed in Figure 12.3. The main findings are as follows:

- Since 1989, Queensland's GS has fallen from 15.7% to 9.9%, implying that although, following the World Bank interpretation of GS, we are possibly on a sustainable growth path, the decline should be of concern to policy makers.

Figure 12.3 Genuine saving estimates for the state of Queensland

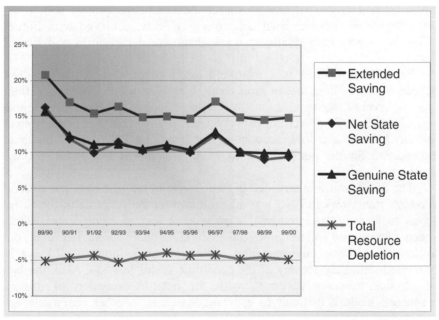

Source: Brown *et al.* (2003).

- Mineral depletion is the leading component of the State's overall depletion of natural resources. It comprises approximately 80% of the total depletion of natural resources.

- The extent of any divergence between saving as measured by State accounts and the GS is dominated by two components – the effects of mineral depletion and ongoing investments in human capital. These two parts of the GS effectively offset each other implying that Queensland's Net State Saving rate as measured in the State Accounts closely approximates the GS.

Case Study 12.2: An Application of Green Accounting to Papua New Guinea[80]

Introduction

The main purpose of this case study was to apply the proposed SEEA in a country at a relatively early stage of development and with as yet moderate environmental problems. One of the challenges of the project was to see what degree of environmental accounting could be achieved with limited effort in country with relatively weak institutional capacities and limited data.

Papua New Guinea is a relatively small country of about 4 million people, 90% of whom live in rural areas. It has few urban centers and there is a low level of industrialization. A few large copper, gold and silver mines contribute about 70 percent of exports. Central government expenditures on the environment are low, accounting for less than 1% of total budgeted expenditure for the period 1986 to 1990, which was the period for this analysis.

The SEEA was implemented in three steps. In the first step, the national accounts framework in PNG was adapted to meet the structure of the SEEA so as to identify environmental expenditures and to incorporate balance sheets of produced and nonproduced (natural) tangible assets. However, due to inadequate data, environmental protection expenditures could not be presented separately for any of the economic agents in PNG. Also, the lack of physical resource accounts, with the notable exception of mineral resources, made it difficult to estimate scarcities in other renewable and nonrenewable resources.

---

[80] The material in the section is based on Bartelmus *et al.* (1993) and Asafu-Adjaye (1991).

In the second step, an attempt was made to value the natural resource scarcities in monetary terms. Two approaches for calculating the depletion costs of the use of scarce natural resources were used—the user cost approach and the net price approach. The user cost approach attempts to convert the stream of revenues from sales of an exhaustible natural resource into a permanent income stream by investing a part of the revenues (referred to as the 'user-cost allowance') over the lifetime of the resource. According to El Serafy (1989), only the remaining amount of the revenues should be considered as 'true income'. The 'net price' is defined as the market price minus all factor costs including a normal return to capital and is the method used in this study. The final step was to compute the discounted future stream of income using the net price. This was used to estimate an 'environmental depletion cost' which was deducted from net value added to obtain estimates of environmentally adjusted net value added of the sector and environmentally adjusted net domestic product of the economy.

Results

Table 12.1 shows estimates of the economic costs of environmental quality degradation of four sectors – agriculture, forestry, mining and energy.

Table 12.1 Papua New Guinea: average annual costs of environmental quality degradation (millions of kina)

Sector	1981–85		1986–91	
	Lower Estimate	Upper Estimate	Lower Estimate	Upper Estimate
Agriculture[a]	9.4	119.0	8.1	119.0
Forestry[b]	10.0	45.0	10.0	61.0
Mining[b]	35.7	101.2	35.7	101.2
Energy[a]	0.03	0.03	0.03	0.03
Total	55.1	265.2	53.8	281.2

Notes: [a]Based on compensation values.
[b]Based on avoidance costs.

Source: Bartelmus *et al.* (1993:125), Table 7-12.

The main environmental impacts of agriculture are from the effects of forest clearing for cultivation. Non-economic, ecological, and related social and spiritual values of forests are lost through the conversion of forests for agriculture and other uses. The main consequences of the depletion of PNG's forests include loss of biodiversity, soil loss, impairment of watershed regulation and nutrient cycles, and the increased risk and intensity of flooding and landslides. The estimates of the costs of forest depletion are for conversion of forest land to agricultural uses. The estimates are based on the rate of compensation to be paid to landowners deprived of traditional uses of the forests by logging activities and should therefore be regarded as very conservative. The low estimates are K9.4 million (1981–85) and K8.1 million (1986–90) per annum. The high estimate is K119.0 million per annum for the period 1981–1991.

As is the case in the agricultural sector, the major environmental impacts of logging are the losses of ecological and cultural functions of forestry resources. The social costs of logging are estimated at about K45 million per annum for 1981–85 and K61 million per annum for 1986–90. In PNG, the main environmental effects of mining are from the discharge of mine tailings that pose a threat to aquatic resources in fresh and marine waters. Estimates of the costs of avoiding environmental damage from mining range from K35.7 million per annum to K101.2 million per annum.

Table 12.2 presents estimates for environmentally adjusted net value added and domestic product using only the lower values of the estimated environmental damage. The results indicate that EDP1 reduces NDP by amounts ranging from 1 to 8 percent and EDP2 by 2 to 10 percent.

Due to the fact that the period under review is short, no definite conclusions can be reached regarding the trends in environmental depletion and degradation. EDP2 (i.e. the additional accounting for environmental quality degradation) reduces NDP by on average a further 2.1 percent. A comparison of the ratios of final consumption (C) to EDP2 indicates that consumption exceeded net (environmentally adjusted) domestic product.

Conclusions

The study shows that it is feasible to implement the SEEA in a developing country situation. But at the same time, it highlights the constraints and limitations of the approach. The main contribution of the study is to reveal the many significant data gaps that exist. The following priorities for environmental data collection were identified:

(i)   It would be necessary to establish natural resource accounts and balance sheets in the areas of forests, soils, subsoil assets of minerals, gas and oil, and fish stocks;

(ii)   There is a need to monitor effluents and loadings of water pollutants from agriculture, mining, industry and municipalities, as well as changes in levels of biodiversity; and

(iii)   Statistics should be developed on environmental (protection) expenditures by the government, industries and households.

Table 12.2 Papua New Guinea: estimates of environmentally adjusted net value added and domestic product, 1986–90 (millions of kina)

Adjusted items	1986	1987	1988	1989	1990
NDP	2313.6	2569.2	2861.6	2698.1	2760.2
EDP1	2186.8	2359.5	2755.3	2672.9	2579.5
EDP2	2132.9	2305.6	2701.6	2619.2	2525.7
[(NDP – EDP1)/NDP]*100	5.5	8.2	3.7	0.9	6.5
[(NDP – EDP2)/NDP]*100	7.8	10.3	5.6	2.9	8.5
(C/EDP2)*100	105.4	105.5	94.5	102.1	109.1

Source: Bartelmus *et al.* (1993:128), Table 7-13.

It is hoped that a strong physical database would enable a more accurate estimate of the environmental costs of economic development to be made. Box 12.1 highlights some of the issues that need to be addressed before the SEEA can be more effectively implemented. These include promotion of awareness and devoting more resources to human resource development and research.

Box 12.1 Green accounting in Papua New Guinea

Papua New Guinea is at an early phase of industrial development and, in general, is yet to experience the kinds of serious environmental problems that we have seen in some countries. The use of an approach such as the SEEA could serve as an 'early warning' system for potentially serious environmental problems. The main constraints to implementing the SEEA in PNG include: (i) awareness; (ii) training; and (iii) research.

Awareness

In spite of the high profile given to environmental issues, individual governments are not putting their money where their mouths are'. In PNG, for example, environmental protection services account for a meagre 0.2 percent of total budget expenditure. There is a need for an environmental awareness campaign aimed at government policy makers and analysts, emphasising the need for effective environmental management.

Training

Statisticians and economic planners in government departments such as Finance and Planning, Agriculture, Minerals and Energy, must be given training not only in the concepts of the SEEA but also in relevant aspects of the emerging discipline of environmental economics.

Research

Considerable research effort is required to provide the necessary inputs into the SEEA. The problem of data availability is especially acute in developing countries such as PNG. Areas of research could include: impacts on agriculture, forestry and fisheries, soil erosion and run-off, and non-market valuation techniques. Research funding could be sought from external agencies such as the World Bank.

Source: Asafu-Adjaye (1991).

Key Terms and Concepts

defensive expenditure
exhaustible resources
gross domestic investment
gross domestic saving
gross state product
human capital
natural capital
net foreign borrowing

net saving
nonmonetary impacts
physical (or produced) capital
physical accounts
pollution damage
resource depletion
transboundary pollutant

Review Questions

1. Explain the objectives of the proposed System of Integrated Environmental-Economic Accounting (SEEA).

2. What are some of the limitations of the SEEA?

3. Define the components of Genuine Saving.

4. How are the following measures of Genuine Saving measured?
   (i) human capital investment;
   (ii) natural resource depletion; and
   (iii) pollution damages.

5. What are the limitations of Genuine Saving as a measure of welfare?

Exercises

1. Why is net saving a better measure of sustainability than gross saving?

2. Critically discuss the World Bank's methodology for measuring genuine (or extended) saving.

3. Give examples of the policy uses of measures of (or extended) genuine saving.

4.   Discuss the practical constraints to implementing the SEEA.

References

Aaheim, A., Nyborg, K. (1995). On the Interpretation and Applicability of a 'Green National Product'. *Review of Income and Wealth*, 41(1):57-71.

Ahmed, Y.A., El Serafy, S. Lutz, E. (eds.) (1989). *Environmental Accounting for Sustainable Development*, The World Bank, Washington, D.C.

Asafu-Adjaye, J. (1991). Environmental Accounting in Papua New Guinea. *Pacific Economic Bulletin,* 8: 42-49.

Bartelmus, P.L.P. (1992). Environmental Accounting and Statistics. *Natural Resource Forum*, February: 77-84.

Bartelmus, P., Lutz, E., and Schweinfest, S. (1993). Integrated Environmental and Economic Accounting: A Case Study for Papua New Guinea. In *Toward Improved Accounting for the Environment*, ed., Ernst Lutz, World Bank, Washington, D.C., pp.108-143.

Bartelmus, P. and van Tongeren, J. (1994). Environmental Accounting: An Operational Perspective. Working Paper 1, ST/ESA, United Nations, New York.

Brown, R., Asafu-Adjaye, J., Draca, M. and Straton, A. (2003). How Useful is the Genuine Savings Rate as a Macroeconomic Sustainability Indicator? Queensland's GSR, 1989-99. Mimeo, School of Economics, the University of Queensland, Brisbane.

Centre for Economic Policy Modelling, CEPM (2002). *Queensland's Genuine Savings Rate, 1989-2000*, Report to the Queensland Environmental Protection Agency. Report prepared by the Centre for Economic Policy Modelling, The University of Queensland.

Dasgupta, P. (1995). Optimal Development and the Idea of Net National Product. In *The Economics of Sustainable Development,* (eds.), I. Goldin and A. Winters, Cambridge University Press, Cambridge.

El Serafy, S. (1989). The Proper Calculation of Income from Depletable Natural Resources. In *Environmental Accounting for Sustainable Development*, (eds.), Y.J. Ahmad, S. El Serafy and E. Lutz, UNDP-World Bank, Washington, D.C.

Hamilton, K. (1994). Green Adjustments to GDP. *Resources Policy*, 20(3):155-168.

Hanley, N. (1997). Macroeconomic Measures of Sustainability: A Survey and a Synthesis. Discussion Papers in Ecological Economics 97/3, Department of Economics, University of Stirling, Stirling, U.K.

Hartwick, J.M. (1990). Natural Resources, National Accounting, and Economic Depreciation. *Journal of Public Economics*, 43:291-304.

Hueting, R., Bosch, P., and de Boer, B. (1992). Methodology for the Calculation of Sustainable National Income. *Statistical Essays*, Central Statistics Bureau, the Netherlands.

Organisation for Economic Cooperation and Development, OECD (2000). *Frameworks to Measure Sustainable Development,* Paris, OECD.

Maler, K.G. (1991). National Accounts and Environmental Resources. *Environmental and Resource Economics*, 1(1):1-16.

Maler, K.G (1995). Welfare Indices and the Environmental Resource Base. In *Principles of Environmental and Resource Economics,* Edward Elgar, Cheltenham.

Pearce, D. and Atkinson, G. (1993). Capital Theory and the Measurement of Sustainable Development: An indicator of Weak Sustainability. *Ecological Economics*, 8(2):103-108.

Repetto, R., Magrath, W., Wells, M., Beer, C., and Rossini, F. (1989) *Wasting Assets, Natural Resources in the National Income Accounts*. World Resources Institute, Washington, D.C.

13. Assessment of Global Environmental Trends and Policy Implications

13.1 Introduction

Over the past century, impressive gains have been made in human development as measured by key socioeconomic indicators such as life expectancy, infant mortality, illiteracy and per capita income growth. However, these gains have been achieved at an environmental cost. The world's population is currently 6 billion and it is expected to increase substantially before stabilising at between 8 and 12 billion by 2050 (UN, 1996). Although global food production is thought to be adequate, there is unequal distribution of food resources and therefore a significant proportion of the population is malnourished.

Global energy use is projected to increase in the future. This implies that, in the absence of any effective policy measures, the buildup of greenhouse gases in the world's atmosphere will accelerate. Acid rain, which was a serious problem in North America and Europe in the 1960s, is now emerging as a serious environmental problem in the Asia-Pacific region. According to Downing *et al.* (1990), 34 million metric tons of SO_2 were emitted in the Asia-Pacific region in 1990, an amount that is 40 percent more than SO_2 emissions in North America in the same period. Areas in the Asia-Pacific region with high acid deposition levels include southeast China, northeast India, Thailand, and the Republic of Korea. These developments pose a threat to crops and ecosystems.

Deforestation continues to be a problem, especially in developing countries. According to the Food and Agriculture Organisation, developing countries lost nearly 200 million hectares of forests between between 1980 and 1995 (FAO, 1997). Allowing for increase in reforestation and forest conservation in the developed countries, the global loss of forest cover is estimated at 180 million hectares between 1980 and 1995, or an average annual loss of 12 million hectares. In 1997, forest fires destroyed vast areas of forestland in Indonesia. It is estimated that between 150,000 to 300,000 hectares of forests were destroyed (EU, 1997). The continued loss of global forest resources poses a serious threat to environmental services such as biodoversity, watershed management, climate regulation, landscape and aesthetic values. Deforestation also threatens the means of livelihood and culture of indigenous people.

Other global resources at risk include water and fisheries resources. Although there are abundant global water supplies to meet the needs of the world's population, the distribution is uneven. Already, parts of the developing world are experiencing varying degrees of water stress due to excessive demands. Given the projections for population growth and industrial development, additional stress is likely to be put on global water supplies. For example, it is estimated that demand for irrigation water will increase from 50 to 100 percent, and industrial water demand will double by 2025 (WMO, 1997). Apart from the problem of water scarcity, there is also the problem of water quality. Rapid industrialisation in developing countries without effective environmental policies will lead to an increase in water pollution. Increase in water pollution affects the quality of groundwater supplies and creates health and sanitation problems.

The projections for fisheries resources are also not favourable. Overfishing of certain fish stocks, which has been a historical problem, has worsened. At present fish stocks in some parts of the world are under threat. For example, fish stocks such as Atlantic cod, haddock, and redfish have nearly collapsed in some parts of the North Atlantic due to overfishing (Grainger and Garcia, 1996). Due to the projected increase in world population, demand for fish is expected to increase. The FAO projects an increase from the current 80 million metric tons to 120 million metric tons by 2010 (FAO, 1997). This level of demand, in the absence of any curbs on overfishing and a significant increase in aquaculture, will be difficult to meet.

13.2   Policy Implications

In spite of the gloomy projections described above, there are some positive signs and there is hope that action can be taken to avert a catastrophe of Malthusian proportions. Following the Earth Summit in Rio de Janeiro, there has been a general consensus that the only way forward is by pursuing a policy of sustainable development. Although, as indicated in the previous chapter, such a policy still needs to be better articulated and translated into concrete action. One hundred and sixty seven nations ratified the 1992 Framework Convention on Climate Change prior to the Earth Summit. In December 1997, these countries met again in Kyoto, Japan to negotiate the first ever legally binding limits on greenhouse gas emissions from developed countries. The Kyoto Protocol aims to cut the combined emissions of greenhouse gases from developed countries by about 5 percent from their 1990 levels by the period 2008–2012 (UN, 1997). The developed countries must agree to contribute specific amounts of emissions reductions toward meeting this target. While the treaty is yet to be ratified, it represents a historic step in international efforts to reduce global emissions.

Another encouraging sign in the advanced countries is the change in consumer attitudes and preferences towards environmentally-friendly products. In supermarkets, consumers are showing a preference for 'green' products, and the message that it pays to be green is slowly filtering through to firms. The rate at which materials are recycled has increased and markets are emerging for recycled products. As discussed earlier, birth rates in the developed countries have declined as a result of increasing employment opportunities for women. Unfortunately, these trends are yet to be experienced in some developing countries. An important factor explaining differences in the levels of environmental awareness in developed countries vis-à-vis developing countries is education. There is therefore a need for formal, as well as, informal programs to educate the general public on environmental issues. Such programs are currently being undertaken by NGOs in some countries, but are in need of government support.

A key question raised in this book is whether there are environmental and resource limits to economic growth. On the basis of the material presented, the answer to this question must be in the affirmative, although this must not be taken as an endorsement of neo-Malthusian predictions about resource use. While technology can allow us to utilise new sources of energy and improve the efficiency of use of existing ones, the law of dimishing marginal productivity suggests that unrestrained economic growth cannot occur. The First and Second Laws of Thermodynamics suggest that unlimited economic expansion is not feasible and that eventually the economy-environment system will reach a steady state. The challenge for sustaining economic growth is therefore to formulate policies for prudent management of our natural resources.

It was demonstrated in Chapter 4 that the market system does not work well for public goods and that it results in overexploitation of resources and excessive pollution. There is therefore a justification for government intervention to correct externalities. To date, government attempts to regulate greenhouse gas emissions, particularly from motor vehicles, have relied on charges (e.g., carbon taxes) command-and-control (CAC) mechanisms, and voluntary agreements (VAs). Carbon taxes have been criticised for their uneven impacts on various socioeconomic groups (OECD, 1995), and it has been argued that they do not guarantee significant reductions in greenhouse gas emissions (Burgess, 1990). An example of a VA in Australia is the National Average Fuel Consumption (NAFC) program which was brokered through a voluntary agreement between the motor vehicle manufacturers and the Australian Federal Government. Wilkenfeld *et al.* (1995) have argued that the NAFC has failed to reduce greenhouse gas emissions in Australia.

Roberts and Spence (1976) have proposed an emissions reduction scheme which combines elements of permits, charges and subsidies. Under this scheme, firms are allowed to trade in pollution permits, but any unused permits can be exchanged for a subsidy. A charge is applied to firms who exceed their emissions quota. Roberts and Spence argue that this mixed approach is superior to the others because even if the initial parameters (e.g. number of permits, levels of subsidy and charges) are wrong, the result would still approximate an optimal solution.

Hensher (1993) has proposed a variation of Roberts and Spence's mixed approach for controlling motor vehicle emissions. The major difference is that his scheme would be applied to consumers rather than to firms. That is, a benchmark level of permissible emissions would be established for each consumer. Consumers who exceed their quota would be subjected to a charge and those who emit below their quota would be subsidised. In addition, trade in emission permits would be allowed. The viability of the scheme depends on the availability of an electronic banking facility that will link point of sale transactions with a central database. While the technology for this sort of scheme currently exists in most advanced countries, there is likely to be public resistance due to concerns about consumer privacy and the likelihood of it being seen as a 'big brother is watching' scheme.

To conclude, let us return to an observation that was made in Chapter 1. That is, in spite of the boom in the global economy over the past thirty years, the benefits of this growth have not been evenly distributed. The advanced countries and, in particular, the newly industrialised countries registered impressive growth rates during this period. However, in many developing countries, the economic growth rate barely kept up with the rate of growth of the population. It was established in Chapters 9 and 10 that there is a significant relationship between economic growth, on the one hand, and economic development indicators and environmental quality, on the other. In view of the lack of sufficient economic growth in some developing countries, and the mixed prospects for the future, there is the possibility that poverty in these countries could increase. Poverty has been shown to be a crucial factor affecting environmental degradation. A growing phenomenon in the last two decades has been urban poverty, which has increased as people migrate into urban centres in search of non-existent employment. The provision of infrastructure support such as housing, water and sanitation services often cannot cope with the rapid influx of people, resulting in increase in environmental pollution.

Poverty alleviation is therefore a major challenge for many governments as we enter a new millennium. However, poverty cannot be eradicated without economic growth. Many developing countries are still dependent on agricultural commodities whose prices remain low or are declining in world markets. At the same time, these countries face mounting national debts. Few have been able to successfully restructure their economies to enhance growth. In the mean time, environmental problems such as deforestation continue to escalate. In order to tackle environmental problems, there is a need for the advanced countries to take concerted action to reduce Third World debt. Global environmental initiatives such as the GEF must be enhanced and made more effective with an injection of adequate capital.

In many countries, there is an urgent need for local and national level policies to provide economic incentives for environmental protection and to transmit correct price signals for the use of natural resources. There is also a need for governments to recognise and safeguard individual and communal property rights. Finally, there is not only a need for firm legislation to support these policies, but also adequate resources to enforce and monitor the legislation.

References

Burgess, J.C. (1990). The Contribution of Efficient Pricing to Reducing Carbon Dioixide Emissions. *Energy Policy*, 18(5): 449-455.

Downing, R., Ramankutty, R. and Shah, J. (1997). *RAINS-ASIA: An Assessment Model for Acid Deposition in Asia*. The World Bank, Washington, D.C.

European Union, EU (1997). *Fires in Indonesia, September 1997*. A report by the EU GIS/Remote Sensing Expert Group to the European Union, European Union, Brussels.

Food and Agriculture Organization of the United Nations, FAO (1996). *The State of World Fisheries and Aquaculture 1996*. FAO, Rome.

Food and Agriculture Organization of the United Nations, FAO (1997). *State of the World's Forests 1997*. FAO, Rome.

Grainger, R. and Garcia, S. (1996). *Chronicles of Marine Fishery Landings (1950-1994): Trend Analysis and Fisheries Potential*. FAO Fisheries Technical Paper, No. 359, Food and Agriculture Organization of the United Nations, Rome.

Hensher, D.A. (1993). *Greenhouse Gas Emissions and the Demand for Urban Passenger Transport: Design of the Overall Approach*. Occasional Paper, No. 108, Bureau of Transport and Communications Economics, Canberra.

Organisation for Economic Co-operation and Development, OECD (1995). *Climate Change, Economic Instruments and Income Distribution*. OECD, Paris.

Roberts, M.J. and Spence, M. (1976). *Effluent Charges and Licences Under Uncertainty*. Journal of Public Economics, 5: 193-208.

United Nations, UN (1996). *World Population Prospects 1950-2050 (The 1996 Revision)*. Population Division, U.N., New York.

United Nations, UN (1997). Kyoto Protocol to the United Nations Framework Convention on Climate Change. Article 3, Annex B, U.N., New York.

Wilkenfeld, G., Hamilton, C., and Saddler, H. (1995). *Australia's Greenhouse Strategy: Can the Future Be Rescued?* Discussion Paper, No. 3, The Australian Institute, Canberra.

World Meteorological Organization, WMO (1997). *Comprehensive Assessment of the Freshwater Resources of the World*. WMO, Geneva.

Useful Internet Resources

1. Climate Ark: www.bapd.org/gcltrk-1.html
    Climate change and alternative energy news and information; research
    materials; links to other sites.

2. Environmental Protection Agency (EPA):
    Global warming: www.epa.gov/globalwarming/
    Ozone layer depletion: www.epa.gov/ozone/
    Cost-benefit analysis:
    www.epa.gov/ebtpages/econcostbenefitanalysis.html

3. Eurostat: www.europa.eu.int/comm/eurostat/
    Statistical office of European Communities; social, economic and
    environmental statistics

4. Food and Agriculture Organisation (FAO): www.fao.org
    Food issues; sustainable development; statistics on agriculture, fisheries,
    forestry, nutrition and economics

5. Friends of the Earth: www.foe.co.uk
    Measuring progress, sustainable development

6. Global Environment Fund (GEF): www.gefweb.org
    Publications, projects funded by GEF

7. International Institute for Sustainable Development (IISD):
    www.iisd.org
    Sustainable development indicators; information on climate change;
    energy, investment; natural resources; trade and economic policy

8. International Monetary Fund (IMF): www.imf.org
    Financial statistics, research reports

9. International Society of Ecological Economists:
    www.ecologicaleconomics.org
    Ecological economics

10. Organisation of Economic Cooperation (OECD): www.oecd.org/ech/
    Trade issues, including trade and the environment

11. Redefining Progress: www.sustainableeconomy.org
    environmental tax reform; sustainability indicators; sustainable
    economics

12. Resources for the Future (RFF): www.rff.org
    Information on environmental management; environment and
    development; natural and biological resources; technology and
    development; public health and environment

13. Sustainable Development Organisations:
    www.webdirectory.com/Sustainable_Development
    Web site hosting for business, organisations (including non-profit/non-
    governmenttal organisations) concerned about sustainable development
    and sustainability.

14. United Nations Development Program (UNDP): www.undp.org
    Human development reports; millenium development goals; water and
    sanitation; environment; gender issues; poverty reduction; energy for
    sustainable development

15. United Nations Environmental Program: (UNEP)
    www.unep.org/unep/eid/geo1/ch/toc.htm
    Global environmental outlook; environmental conventions; GEF
    coordination; regional cooperation; environmental policy development

16. United Nations Framework Convention on Climate Change (UNFCC):
    www.unfcc.de/
    Cimate change

17. United Nations Population Fund (UNPF):
    www.un.org/esa/population/unpop.htm
    Population

18. World Bank: www.worldbank.org/environmetaleconomics
    Environmental economics; environmental indicators; green accounting

19. World Health Organisation (WHO): www.who.org
    Health issues

20. World Resources Institute (WRI): www.wri.org

Climate change www.wri.org/climate
Agriculture and food; biodiversity; climate change
and energy; coastal and marine systems; water resources;
forest, grasslands and drylands; institutions

21. World Summit on Sustainable Development (WSSD):
www.johannesburgsummit.org/
Sustainable development

22. World Trade Organisation (WTO): www.wto.org
International trade issues and environmental quality

23. World Widlife Fund (WWF): www.panda.org/
Conservation of species, forests, freshwater and
marine resources; climate change; environmental policy issues

24. Zero Population Growth: www.populationconnection.org/

Answers to Numerical Exercises

Chapter 3

Question 3

(a) At equilibrium, demand = supply, therefore the equilibrium price clears the market. Therefore equating the two equations, we have

$$6q = 180 - 4q$$

where q is the equilibrium quantity. Solving for q gives a value of 18 units. Substituting q=18 into either the supply or demand function gives an equilibrium price of $108.

(b) At a price of $40 per unit, 35 units will be demanded but only 7 units will be supplied and therefore there will be an excess demand (shortage) of 28 units.

(c) The new supply equation is $p_{TAX} = 6q^s + 20$, giving $p^* = 116$, $q^* = 16$.

(d) If the government imposes a price ceiling of $48 per unit, 33 units will be demanded but only 8 units will be supplied and therefore there will be a shortage of 25 units on the market.

Question 6

(a) The new supply equation will be $p_{TAX} = 6.6q^s$, giving $p^* = 112$, $q^* = 17$. Therefore the consumer surplus will be given by:

$$\frac{1}{2} \times (\$180 - \$112) \times 17 = \$578$$

(b) The producer surplus will now be ½ × ($112 – $0) × 17 = $952.

(c) The net benefits after the tax policy will be the sum of consumer surplus (a) and producer surplus (b): $578 + $952 = $1530.

(e) The deadweight loss is given by difference between the net benefits under the pre-tax scenario minus the net benefits under the post tax scenario.

The net benefits pre-tax = consumer surplus + producer surplus
$$= [½ × ($180 – $108) × 18] + [½ × $108) × 18]$$
$$= $648 + $972$$
$$= $1620$$

The deadweight loss is therefore given by $1620 – $1530 = $90.

Chapter 5

Question 1
(a) The value of houses on the Southside is $3.15 billion, and $1 billion on the Northside.
(b) There is a net gain of $900 million on the Southside and a net loss of $500 million on the Northside, giving an overall net gain of $400 million.

Question 2
(b) The consumer surplus for a single visit per person is given by:

$$½ × (140 – 120) × 1 = $10$$

(c) The total consumer surplus for an average visit (i.e. 5 visits per annum) is 5 × $10 = $50.

(d) The aggregate consumer surplus per annum for the national park is given by: $50 × 50,000 = $2.5 million.

Note: the above analysis assumes there is no entry fee.

Question 3

(a) If the project does not go ahead the average farm price will be given by:
Farm price = 500000 + 30(200) − 100(15) + 2000(0) = \$504,500.

(b) If the project goes ahead the average farm price will be given by:
Farm price = 500000 + 30(200) − 100(15) + 2000(1) = \$506,500.

(b) The change in average farm price if the project goes is \$506,500-\$504,500 = \$2000. This result could also have been obtained by differentiating the hedonic price equation with respect to IRRIG.

Question 4

Given an entry fee of \$1, demand for visits will be 5500 persons. The consumer surplus will be given by:

$$\frac{1}{2} \times (\$4 - \$1) \times 5500 = \$8250$$

If the entry fee is increased to \$3, demand will fall to 2000 and the consumer surplus will be:

$$\frac{1}{2} \times (\$4 - \$3) \times 2000 = \$1000$$

The loss in consumer surplus is given by: \$8250 − \$1000 = \$7250.

Chapter 6

Question 1

The NPV at a 10% discount rate for Project A is \$324,070 and that of Project B is \$779,010. Therefore, Project B is more profitable.

Question 2

For Project A, the present value of costs, using a 10% discount rate is \$867.77 and the present value of total benefits is \$2073.64. Therefore the cost-benefit ratio is 2073.64/867.77 = 2.39.

For Project B, the present value of costs, using a 10% discount rate is \$1735.54 and the present value of total benefits is \$2955.45. Therefore the benefit-cost ratio is 2955.45/1735.5401 = 1.70.

The rankings are not the same as in (1). Comparing just the BCRs, we will rank Project A above B because it has a higher BCR.

Question 3

The IRR for Project A is 15%, and 18% for Project B. On the basis of the IRRs, Project B is preferred to Project A.

Question 4

The NPV and IRR provide the same ranking, while the BCR provides a diferent ranking to the other two. The BCR is not reliable for comparing projects of different sizes and is not suitable for ranking projects.

Question 5

The sensitivity analysis of the effects of changes in benefits and costs on the estimated IRR is as follows:

		Change in net economic benefits				
		−40%	−20%	0%	+20%	+40%
Change in total costs	−40%	18	26	33	39	45
	−20%	10	18	24	30	35
	0%	5	12	18	23	28
	+20%	1	8	13	18	22
	+40%	−3	4	9	14	18

The estimated IRR is fairly robust to small changes in benefits and costs. For example, a 20% reduction in the estimated net economic benefits, with no changes in costs reduces the IRR to 12%, which is above the discount rate of 10%. A 20% increase in costs (with benefits unchanged) reduces the IRR to 13%, which is above the discount rate of 10%.

Question 6

(a) On the basis of just the NPVs, Option 2 — Pollute and clean up would be preferred.

(b) If there were a chance of irreversible damage to the environment, Option 1 — Prevent pollution would be recommended. This is because the cost of irreversible damage would be very high and likely to off-set the total benefits of the pollute and clean up option.

Index